Studies in Brain and Mind

Volume 11

More information about this series at http://www.springer.com/series/6540

Nada Gligorov

Neuroethics and the Scientific Revision of Common Sense

 Springer

Nada Gligorov
Department of Medical Education
Icahn School of Medicine at Mount Sinai
New York, NY, USA

ISSN 1573-4536   ISSN 2468-399X   (electronic)
Studies in Brain and Mind
ISBN 978-94-024-0964-2   ISBN 978-94-024-0965-9   (eBook)
DOI 10.1007/978-94-024-0965-9

Library of Congress Control Number: 2016951491

Printed on acid-free paper

This Springer imprint is published by Springer Nature
The registered company is Springer Science+Business Media B.V. Dordrecht

To my grandma, for whom I was named
For her intelligence, her humor, and her
strength

Acknowledgments

I began planning this book several years ago, hoping to combine my two main areas of interest in philosophy: bioethics and philosophy of mind. The process of writing was at times challenging, and many people helped nudge me along the way. In particular, I would like to thank Robert Baker and Rosamond Rhodes for their support and their encouragement at the outset of this project. I would also like to thank Abraham Schwab for reading the entire manuscript and for his helpful comments. I am also indebted to an anonymous reviewer for many comments that helped substantially improve the chapters and the main theme of the book. I am incredibly grateful to my mother-in-law, Rose Krieger, who carefully read two versions of my manuscript and made numerous comments that helped me improve the readability of the book. I also wish to thank my family: my father, Vladimir Gligorov, for his steady encouragement and for always knowing what I mean; my mother, Bosiljka Stamenovic, and my stepfather, Svetislav Stamenovic, for their unabated confidence in me; and my father-in-law, Joseph Krieger, for his love and support. Finally, I would like to thank my husband, Stephen Krieger, who changed my life and made every effort worthwhile.

The ideas and arguments presented in this book evolved over several years, and sections of this book were already published elsewhere. Section 6.2 in Chap. 6 is based on a previously published manuscript by Gligorov and Krieger (2010), titled "Functional Brain Imaging, Free Will, and Privacy" in Stefane M. Kabene, ed., *Healthcare and the Effect of Technology: Developments, Challenges, and Advancements*, IGI Global Publishing. In Chap. 8, Sect. 8.4 is a modified version of "A Defense of Brain Death" published in *Neuroethics*, Volume 9(2): 119–127.

Contents

Chapter 1
Introduction

Abstract Neuroethics is an emerging interdisciplinary field with unsettled boundaries. Many of the ethical issues within the purview of neuroethics could be
described as resulting from the clash between the scientific perspective on concepts
such as free will, personal identity, consciousness, etc., and the putatively commonsense conceptions of those terms. The assumption that undergirds the framing of the
conflict between these two approaches is that advances in neuroscience, psychiatry,
and psychology can be used to explain phenomena covered by commonsense concepts and in some cases undermine them entirely. This book is focused on the examination of the particular relationship between developments in neuroscience and
commonsense moral concepts. Common sense, I argue, has been misinterpreted as
a static, either foundational or degenerative, basis of our morality, when it is an ever-
shifting repository of theories from many domains. Within this discussion, I focus
on the application of neuroscience to human beings, i.e., the ethics of neuroscience.
But I also cover issues within the purview of the neuroscience of ethics, and attempt
to address the infiltration of neuroscientific knowledge into everyday parlance and
the impact of that on our commonsense morality and psychology.

1.1 Introduction

Neuroethics is an emerging interdisciplinary field with unsettled boundaries. Some
have defined neuroethics most broadly to designate ethical issues pertinent to information about the brain (Glannon 2007, p. 4). Adina Roskies (2002) has distinguished two separate areas of inquiry within neuroethics, i.e., the *ethics of
neuroscience* and the *neuroscience of ethics*.

Ethics of neuroscience overlaps with both research and clinical ethics. For example, whether it is permissible to do studies with participants who suffer from neuro-
degenerative disease such as Alzheimer's, as well as how such research could be
ethically conducted, is a research ethics issue as it applies to neuroscience. The
ethics of neuroscience also includes the application of neuroscientific knowledge to
human beings. In this way, neuroethics overlaps with clinical ethics.

Clinical ethics includes the application of criteria for brain death, as well as the
very concept of brain death. Additionally, the off-label use of psychopharmacological agents for cognitive enhancement, memory modification, and even their use to

© Springer Science+Business Media B.V. Dordrecht 2016   1
N. Gligorov, *Neuroethics and the Scientific Revision of Common Sense*,
Studies in Brain and Mind 11, DOI 10.1007/978-94-024-0965-9_1

alter undesirable personality traits, such as shyness or introversion, are all trailed by significant ethical considerations.

Neuroscience of ethics is a field focused on examining the neural underpinnings of moral behavior. Researchers in this field investigate the neural realization of traits that are often deemed necessary for moral behavior, such as free will, autonomy, and even personal identity. Some researchers have examined the role of emotions in our moral reasoning, thereby questioning the predominant view that morality is the result of rational deliberation (Greene et al. 2001). More generally, evidence from neuroscience has been used to adjudicate even long-standing disputes in moral theory, such as the one about consequentialism and deontology (Greene 2008).

This book is focused on the examination of the particular relationship between developments in neuroscience and commonsense concepts, such as free will, personal identity, privacy, etc., which feature prominently in moral discourse. Common sense, I will argue, has been misinterpreted as a static, either foundational or degenerative, basis of our morality, when it is an ever-shifting repository of theories from many domains. Within this discussion, I will focus on the application of neuroscience to human beings, i.e., the ethics of neuroscience. But I will also cover issues within the purview of the neuroscience of ethics, and attempt to address the infiltration of neuroscientific knowledge into everyday parlance and the impact of that on our commonsense morality and psychology.

1.2   Issues in Neuroethics

There are a number of already established topics within neuroethics. In this section, I will describe some of those topics, without yet challenging their formulation or interpretation of their ethical significance. The remainder of the book aims to rethink a number of the issues mentioned in this section.

An often tackled neuroethics topic is the ethical evaluation of the use of brain-imaging technology, including computed tomography (CT), single photon emission tomography (PET), magnetic resonance imaging (MRI), single photon emission computed tomography (SPECT), and functional magnetic resonance imaging (fMRI). Functional MRI is a much discussed imaging technique because it has enabled researchers to argue that there are discrete pockets of brain function that correlate with specific cognitive abilities, such as memory or learning, and in some cases even correlate with particular thoughts (Kay et al. 2008). Studies featuring fMRI images have also captured the imagination of the public because the colorful images facilitate the inference that fMRI captures activity in the brain as it happens.[1]

---

[1] The seductiveness of brain imaging was illustrated in the study by Weisberg et al. (2008). The participants in this study were given a number of different explanations of particular psychological phenomena, including explanations that were designed to be of notably poor quality. The participants in the study consisted of three groups: neuroscientists, neuroscience students, and lay adults.

The moral evaluation of the use of fMRI and other types of brain imaging centers on the question of whether it is a morally disruptive technology, the use of which could compel us to revise some of our current norms and practices. There are reasons to think that brain imaging is that kind of technology. Within the ethics of neuroscience, brain imaging has been seen as morally disruptive because it could challenge notions of privacy. There are several specific cases in which these issues have been raised. One of the more widely discussed future uses of functional neuroimaging technology (including fMRI) is for lie detection. The technique of fMRI lie detection is based on the functional imaging of the cognitive process required for deception. The basis of fMRI lie detection is the finding that more activation is seen in the prefrontal and anterior cingulate regions in the "lie condition" relative to the "truth condition" in an experimental setting where participants were asked to either lie or tell the truth (Fenton et al. 2009). Other fMRI studies have shown that there is a difference between the brain activity when a person is in a familiar setting, room, or part of town, and the brain activity when a person is in a new environment (Meegan 2008), which introduces the possibility that fMRI could be used for placing suspects at the scene of a crime. More broadly, the possibility that brain imaging can be used to infer what a person is thinking raises the worry about mind reading (Meegan 2008). Functional MRI studies have also been used to identify brain activity correlated with the experience of pain (Coghill et al. 2003). Given the correlation between pain experience and particular kinds of brain activation, brain imaging might be used as a tool to distinguish between true and false reports of pain, which could help identify patients who are malingering (Kolber 2007).

Additionally, brain imaging has been cited as having the potential to facilitate diagnosis of mental illness before it becomes symptomatic, for example, if fMRI could be used to identify abnormal brain functioning even before a person starts exhibiting behavior associated with a disease such as schizophrenia (Fenton et al. 2009). Although the use of brain imaging as a diagnostic tool would have some obvious medical benefits, such as early diagnosis coupled with timely treatment, there is the potential for negative consequence as well. It might stigmatize individuals with the early diagnosis and prevent them from becoming full members of society.

Within neuroscience of ethics, when brain imaging is taken to contribute to the overall project of reducing our psychology to neuroscience, it has been interpreted as undermining assumptions about people's abilities to act morally. For example, Greene and Cohen (2004) argue that the concept of free will ought to be eliminated because the concept is not supported by growing evidence about how the brain functions (Greene and Cohen 2004). Alternatively, Kaposy (2010) argues that we do not have an obligation to revise our moral concepts based on scientific evidence because many of our social structures function on the presumption that persons are rational, autonomous, and have free will.

The study found that all three groups did well at identifying the poor explanations, except when those explanations were preceded with the words "Brain scans indicate...." Although this did not sway the neuroscientists, the students and lay adults were more likely to accept the bad explanation.

An additional emerging topic in neuroethics is the consequences of memory manipulation. There are several pharmacological ways in which human memory can be affected. Studies have shown that drugs such as scopolamine (Squire and Caine 1980), benzodiazapines (King 1992), and kinase inhibitors (Serrano and Pastalkova 2006) can be used to disrupt the consolidation of memories and prevent the transition of memories from short- to long-term memory. Another often-discussed way in which memory can be manipulated is through the use of beta-blockers, specifically propranolol. Beta-blockers already have a number of medical uses for a variety of cardiovascular conditions, including hypertension. They can also be used to diminish or block the cardiovascular response caused by the neurotransmitters adrenaline and noradrenalin.

In a study conducted by Pitman et al. (2002) of patients who were in a traumatic accident, those who were given beta-blockers were less likely to experience post-traumatic stress disorder (PTSD). One interpretation of why beta-blockers would have this effect is that by dampening stress reaction in the body, they diminish the emotional impact of the memory of the traumatic event. The person remembers the event, but the event fails to have the appropriate emotional valance. It seems, then, that reducing the emotional reaction to a traumatic event can affect the memory of it (Glannon 2007 48).

Assuming that beta-blockers could be efficacious in treating PTSD, they could be especially beneficial if they could be used to curb the incidence of PTSD in soldiers returning from combat and adjusting to civilian life. Nonetheless, there are several ethical issues that are raised in connection with their use for memory modification (Kass 2003; Liao and Sandberg 2008; Henry et al. 2007; Wasserman 2004). According to those who endorse the view that personal identity depends on the maintenance and continuation of a number of important memories, disruption in memory could cause a discontinuity in personal identity. For example, traumatic experiences are sometimes foundational to personal identity (Erler 2011). Beta-blockers could perhaps abate their formative impact by lessening the traumatic properties of the memory.

Memory modification could have additional consequences, especially if there is a moral obligation to have appropriate moral reactions (Liao and Sandberg 2008; Kass 2003). For example, consider a soldier who during combat shoots and kills somebody. Even in cases where the soldier believed that the killing was justified, the traumatic impact of this violent event could take a psychological toll. One might regret such events and even feel remorse. Beta-blockers could help minimize the psychological impact of the traumatic event. According to Liao and Sandberg (2008), there is an obligation to remember bad memories. For those who have committed crimes or done something morally reprehensible, the bad memory of the event, including the memory of the bad consequences, can be instrumental. They might promote appropriate remorse and become a deterrent in the future. Furthermore, the emotional impact of the memories could shape a person's moral character, and the negative impact of bad memories might become instrumental in improving his or her character.

Innovative uses of pharmacology have also been discussed in relation to medical enhancement. Medical enhancement has been defined by commentators as the use of medical intervention aimed at the improvement of normal individuals. There are different types of purported enhancers, some surgical and others pharmacological. A frequently discussed purported cognitive enhancer is methylphenidate (marketed in the United States as Ritalin®). For individuals with attention deficit hyperactivity disorder (ADHD), methylphenidate is used to treat that condition. In normal individuals, studies support the claim that methylphenidate can increase concentration and improve performance on cognitive tasks (Elliot et al. 1997; Mehta et al. 2000; Tye et al. 2010). Another potential cognitive enhancer is modafinil (Provigil®). It has been FDA approved for the treatment of narcolepsy, but has been prescribed off label for a variety of sleep disorders, including sleep apnea. Studies on normal healthy adults who do not suffer from sleep disorders showed that modafinil could be successfully used to counteract the negative effects of sleep deprivation (Grady et al. 2010; Sugden et al. 2012).

Cognitive enhancement and, more broadly, medical enhancement have been interpreted as expanding the purpose of medicine. Commentators such as Sandel (2004) and Fukuyama (2002) have argued that such extension of the purpose of medicine is not justified. They argue that there is a distinction between the use of medicine for treatment and the use of it for enhancement. In turn, that distinction can be used to parse the uses of medicine that are morally justified from those that are not. For example, the use of Ritalin for the treatment of ADHD would be justified, but its use for cognitive enhancement of normal individuals would not be morally justified. Alternatively, Daniels (2000) and Synofzik (2009) argue that the distinction between treatment and enhancement is not easy to draw and that some forms of enhancement are morally justified.

The use of cognitive enhancers could also exacerbate the already difficult task of the just allocation of resources and opportunity, which was noted in Farah et al. (2004). If the use of cognitive enhancers becomes prevalent, their use might increase the averages necessary for scholastic achievement. Thus, those who might not be able to obtain cognitive enhancers might become permanent underperformers and become unable to compete at a level necessary to perform well academically and have a successful life.

Advances in brain science have also challenged our conception of death by introducing the concept of brain death. The current criterion for the diagnosis of brain death requires the death of the entire brain, including the brain stem. In order to diagnose brain death, a qualified physician, usually a neurologist, is required to go through the brain-death protocol--a set of diagnostic tests. A diagnosis of brain death is in effect a determination that the whole brain has died, including the brain stem. With artificial ventilation, however, the flow of oxygen in and out of the lungs can be continued and a brain-dead individual can maintain a heartbeat for a significant amount of time (Shewmon 1998).

The brain-death criterion is a legally supported way of diagnosing death in most of the United States, but the legal sequelae of a diagnosis of death can differ between cardiac and brain death. For example, hospitals in New York and New Jersey are required by law to provide reasonable accommodations for those families who for religious or moral reasons do not recognize brain death as death. Such accommodations could include continuing ventilation, nutrition, and hydration for the individual, as well as providing medication for some short period of time. Beyond religious and cultural considerations, some have challenged brain death because it does not fit the traditional criteria associated with cardiac death (Collins 2010; Shewmon 2010).

A more radical reconceptualization of the concept of death is put forth by proponents of the "higher-brain" death criterion (Glannon 2007; Veatch 2005). Proponents argue that the death of the individual corresponds with the death of the person, the existence of which requires the ability for mental functioning, including the ability to be conscious. Given that mental functioning and consciousness are thought to be realized in the cerebral cortex, the death of that organ would signal the loss of personhood. And when the person dies, the individual has died. Based on the cardiac criterion of death, a brain-dead individual would be considered alive, while by the higher-brain criterion for death, some individuals who are in vegetative states and are currently considered alive would be counted as dead (Glannon 2007).

Moral and other philosophical issues concerning altered states of consciousness, such as vegetative states and minimally conscious states, have also become part of the neuroethics canon. A number of studies indicate that individuals diagnosed as being vegetative still retained cortical activity (Owen et al. 2006; Owen and Coleman 2008; Monti et al. 2010, 2013). Brain imaging studies have shown that the clinical diagnosis of vegetative state is not enough to determine whether an individual has a complete loss of function of the cerebral cortex. Moreover, there is evidence that being in a vegetative state is not entirely irreversible and that treatment alternatives might be on the horizon (Du et al. 2014; Machado et al. 2014). In light of these findings, there are a number of ethical implications to consider, including the just allocation of medical resources and what we owe individuals in vegetative states, as well as the issue raised by Glannon (2007), whether to classify such individuals as persons.

1.3   The Scientific Revision of Common Sense

Many of the ethical issues within the purview of neuroethics could be described as resulting from the clash between the scientific perspective on concepts such as free will, personal identity, consciousness, etc., and the putatively commonsense conceptions of those terms. The assumption that undergirds the framing of the conflict between these two approaches is that advances in neuroscience, psychiatry, and psychology can be used to explain phenomena covered by those commonsense concepts and in some cases undermine them entirely.

Commonsense and scientific conceptions are particularly contrasted in debates about free will. For example, common conceptions of free will are said to be incompatible with neuroscientific discoveries that many of our actions are not the result of conscious decisions. This in turn could have implications for moral responsibility, especially if it is also the case that moral responsibility requires conscious willing.

The characterization of psychological states is an additional domain where commonsense is contrasted with science. For example, use of brain imaging has been said to challenge the folk-psychological conception of mental states as inherently private (Richmond 2012). This ostensibly folk-psychological notion of inherent privacy is then used to determine the permissibility of brain imaging technology. Also, the subjectivity of mental states, such as pain, is often credited to commonsense psychology and compared with the scientific demands for objective characterization of that phenomenon. The subjectivity of pain states is often used to support the primacy of verbal reports of pain to determine the presence and character of the pain an individual is experiencing. This in turn contrasts with the objectivity required for the scientific study of pain states, which seeks to explain how the same noxious stimuli might produce different experiences of pain, and even how in the absence of any physical damage, an individual might still report feeling pain.

Similarly, disruptions in personal identity have been cited as a potential risk of the use of cognitive enhancers and memory modifiers, especially as they might violate common conceptions of acceptable modes of psychological change. Because pharmacological means of changing oneself differ from accepted, more gradual attempts to change the self, they are judged as unwarranted. Even the purportedly traditional notion of death is said to be challenged by the scientifically laden conception of whole-brain death. This has led some to question whether brain death properly belongs to the concept of death.

I argue that the contrast between these commonsense and scientific conceptions is generated by the faulty characterization of common sense, which then affects both the evaluation of commonsense concepts and the formulation of the ethical issues that arise from this contrasting of commonsense moral concepts with scientific ones. To propose what I think is the most apt way to characterize commonsense concepts, I utilize positions already developed in philosophy of mind about the nature of folk or commonsense concepts. Specifically, I adopt the view that folk morality is, like folk psychology, an empirically evaluable theory. Further, I utilize David Lewis's method (1972) for circumscribing the boundaries of a commonsense view. Like folk psychology, folk morality can be characterized by collecting platitudes used in everyday parlance that feature terms like free will, personal identity, privacy, etc. The collection of the relevant platitudes implicitly defines those concepts by specifying their causal roles.

Adopting this view about the nature and scope of common sense then allows me to reexamine the relationship and seeming conflicts between commonsense and scientific concepts about the moral domain. When addressing the challenge to folk morality, I argue that the boundaries of common sense are hard to circumscribe. I argue for this in three ways: First, I argue that that the utilization of folk moral concepts requires the endorsement of an empirically evaluable theory in the same way as the endorsement of scientific conceptual frameworks requires the endorsement of an empirically evaluable theory. Second, I argue that commonsense concepts have changed over time and that some of those changes were due to the incorporation of scientific facts into our everyday parlance. Third, I argue that using Lewis's method of collecting platitudes to arrive at an implicit functional definition of the theoretical terms of a commonsense theory (either mental or moral) gives us only a sample of current common sense, one that is not representative of common sense over time or across cultures. Using these three arguments, I conclude that unqualified eliminativist claims, for example that neuroscience is falsifying the common notion of free will, are not substantiated.

Despite my anti-eliminativist stance, I also argue that commonsense conceptions should not be accorded a privileged status. When discussing pain, for example, I argue that the characterization of pain as an entirely subjective phenomenon should be challenged even if that tenet is part of commonsense conceptions of pain. Furthermore, I argue that in some cases what is described as a commonsense concept, say, the traditional conception of death, is the result of the seamless incorporation of scientific developments into everyday parlance. For example, the cardiac criterion for death is often thought to be in line with a traditional conception of death, while brain death is considered a scientifically laden concept. This is then used to argue that brain death is not death in the same sense. But it is of note that the cardiac criterion of death was preceded by an alternative criterion, the criterion of putrefaction, and that the establishment of the cardio-pulmonary criterion of death trails the invention of the stethoscope, as brain death was established after the introduction of the medical respirator.

When addressing the changes to commonsense concepts, I argue that developments in neuroscience are often continuous with past scientific developments and although they might affect aspects of our commonsense conceptions, they will not challenge them in entirely novel ways. This argument is applied in chapters where I discuss mental privacy and personal identity. For mental privacy, I argue that brain imaging developments will not challenge our notions of mental privacy, but might expand our view of bodily privacy to include information about our brains. Furthermore, I argue that the concept of informational privacy already covers facts about individuals regardless of how they were obtained, especially if their unwarranted release could be harmful.

Similarly, I argue that although neurocognitive enhancers or memory modifiers could produce psychological changes in an individual and even result in a changed personality, their use is permissible. Given that psychological change occurs over time in a variety of different ways, through education, religious conversion, or personal experience, then pharmacological means of psychological change should also be permissible because they are not relevantly dissimilar to those more accepted ways of altering one's personality. In sum, this book addresses the relationship between commonsense and scientific concepts in two ways--some chapters are focused on the scientific challenge to ostensibly commonsense moral concepts and others highlight how biomedical developments have changed, and will continue to change, aspects of our folk morality.

Because issues in ethics of neuroscience and neuroscience of ethics depend on the connection drawn between philosophical approaches to morality and the human psychological capacities that underlie our ability to reason and act morally, a secondary strategy adopted in this book is to establish a link between philosophical approaches to the nature of mental states, or more generally psychological states, with debates in neuroethics and to assess the impact of the former on the latter. I draw this connection at the very outset of the book, when I settle on a view of commonsense concepts and emphasize it throughout many of the chapters, including chapters on brain imaging, pain, and free will, and even when discussing brain death.

1.4   Chapter Overview

My approach in this book is to examine the relationship between commonsense concepts and scientific findings that might give rise to alternative conceptual frameworks. The view presented in Chap. 2 is the basis for the rest of the book and grounds the more specialized arguments presented in Chap. 3 and beyond.

In Chap. 2, I endorse functionalism about folk-psychological mental terms and apply that view to folk morality. I then describe the purported incompatibility between commonsense and scientific concepts about mind. Although I accept that any conceptual framework could be in principle revised, I provide reasons for why I disagree with arguments for the elimination of commonsense concepts. Nonetheless, I do not think that commonsense concepts are privileged nor do I require that scientific frameworks account for those concepts. Instead, I argue that there is not a principled distinction between commonsense concepts and scientific ones. The concepts we use in everyday life change over time and are influenced by science. In this way, I deflate the purported discontinuity between commonsense and scientific conceptual frameworks as they feature in discussions of psychology or morality.

In Chap. 3, I discuss the purported incompatibility between free will and the evidence for unconscious volitional actions. I begin by presenting a study by Benjamin Libet that has been interpreted to be empirical evidence against the concept of free will. Libet formulates his view of the "common notion" of free will, which requires consciousness (Libet 1999). Assuming this kind of conception of free will, scientific findings of volitional action without consciousness are taken to show that the purportedly common notion of free will is erroneous. I challenge the view that our commonsense notion of free will has been properly characterized by Libet (1999). Additionally, I argue against the view that our commonsense concept of free will is strictly tied to consciousness.

In Chap. 4, I focus on cognitive enhancement. More specifically, I assess the claim that personal identity will be threatened for those individuals who choose to use cognitive enhancers. In arguing against this claim, I distinguish between two different notions of identity, numerical identity and narrative identity. I argue that numerical identity as it pertains to the philosophical puzzle of establishing that an individual is one and the same over time is not relevant to the discussion of the permissibility of the use of cognitive enhancers. In contrast, narrative identity--defined as each person's story of who he or she is, including such individual characteristics as one's preferences, values, and personality traits--is a suitable notion for the evaluation of the use of cognitive enhancers. However, arguments that the use of cognitive enhancers will disrupt personal identity often tacitly confuse numerical and narrative identity. I argue that narrative identity is a first-person, subjective attempt to formulate identity and that it does not have the normative force of numerical identity. Because narrative identity cannot support objective criteria that would allow for the categorization of certain types of changes in self as impermissible, I argue, there is no reason to think that changes obtained through cognitive enhancement are less morally justified than other types of changes in self.

In Chap. 5, I discuss issues related to memory manipulation, including memory enhancement, erasure, or modification. The ethical objections to those are similar to objections raised against cognitive enhancers, as changes in memory might cause discontinuity in personal identity. I present the current state of affairs with regard to memory modification in order to qualify the discussion about the possibility of memory enhancement, erasure, and modification. I address the claims that memory is in some way fundamental to our sense of self by arguing that memory is better characterized as a reconstruction of past events than as a veridical record of the past. Similarly, I argue that our memory of our past selves and our past actions is flawed and does not likely serve as the primary basis for a sense of self or help to maintain a sense of continuity. Because of that, I do not think that changes in memory, whether natural or pharmacological, cause discontinuity of self. My argument extends also to the claim that authenticity can be challenged through memory modification. I reprise my argument that narrative identity is not normative and argue against the view that authenticity can be determined using objective criteria.

In Chap. 6, I evaluate the argument that the use of fMRI poses a threat to mental privacy and I challenge the argument that this type of privacy requires extra protections. I begin by reviewing all the positions about the nature of mental states that establish a category of mental privacy and conclude that none of those views can support both the claim that there is a category of mental privacy and that this type of privacy can be violated through the use of brain imaging. This is because views that characterize mental states as subjective also maintain that the subjective aspects of mental states are only accessible through introspection. I further argue that the only position about the nature of mental states that erases the epistemological gap between introspection and third-person access to our inner states (in this case, brain states or processes) is eliminative materialism. Eliminativism, however, does this by denying the categories of folk psychology, including the category of mental privacy. Finally, I argue that because no view about the nature of mental states can support the argument that 'brain reading' will result in 'mindreading,' fMRI does not and will not pose a threat to mental privacy. I conclude that special protections for mental privacy are not required because an existing concept of informational privacy already protects, at least in principle, the privacy of information about patients and about research participants in whatever way it is obtained.

In Chap. 7, I focus on the issue of the subjectivity of pain. I distinguish among the various ways in which pain is considered to be a subjective phenomenon, including introspectability, privacy, and incorrigibility. I argue that introspectability and privacy are features that could be shared by states both mental and physical. The kind of subjectivity that is often thought to threaten the scientific study of pain arises only when introspectability and privacy of inner states are coupled with a theory of pain states that posits nonphysical properties to account for the content of pain. I also argue that pain is not incorrigible. I use aspects of the argument presented in Chap. 2 to claim that the first-person identification of pain states requires the possession of a rudimentary conceptual framework that includes the concept of pain. This conceptual framework changes over time as an individual is exposed to a variety of different noxious stimuli and acquires a wider vocabulary to express the feeling of pain. Given that the identification of pain requires a concept of pain and that changes in the relevant conceptual framework can alter the feeling of pain, I argue that pain is not incorrigible. Although I acknowledge that there are currently no established criteria to challenge or circumvent a first-person report of pain, there are promising new strategies that could lead to the development of such a tool.

In Chap. 8, I defend the whole-brain criterion of death. I argue that death is not a commonsense concept: All the properties attributed to death stem from its role in a biological theory about the functioning of a human organism. I contend that a biological theory can establish a physical state as the moment of death, but I show using historical examples that the identification of the physical moment of death can change over time as our theories about human biological function are modified. I maintain that a definition of death should not focus only on somatic integration and that the body and brain dualism pervasive in the brain-death literature should be rejected. Instead, the conception of the cessation of functioning of the organism as a whole should apply to the functioning of both the body and the brain. The functioning of the organism as a whole has three major elements: integrated psychological functioning, including memory, consciousness, emotional processing; the integration of psychological and physical processes, such as running in fear; and the integrative functions of the body. I argue that this integrated functioning of the organism as a whole, reconceived to include those three elements, can be used to support brain death as a criterion of the death of the organism.

References

Coghill, R., McHaffie, J. G., & Yen, Y. F. (2003). Neural correlates of interindividual differences in the subjective experience of pain. *Proceedings of the National Academy of Sciences, 100*(14), 8538–8542.

Collins, M. (2010). Reevaluating the dead donor rule. *Journal of Medicine and Philosophy, 35*(2), 1–26.

Daniels, N. (2000). Normal functioning and the treatment–enhancement distinction. *Cambridge Quarterly of Healthcare Ethics, 9,* 309–322.

Du, B., Shan, A., Zhang, Y., Zhong, X., Chen, D., & Cai, K. (2014). Zolpidem arouses patients in vegetative state after brain injury: quantitative evaluation and indications. *The American Journal of the Medical Sciences, 347*(3), 178–182.

Elliot, R., Sahakian, B. J., Matthews, K., Bannerjea, A., Rimmer, J., & Robbins, T. W. (1997). Effects of methylphenidate on spatial working memory and planning in healthy young adults. *Psychopharmacology, 131,* 196–206.

Erler, A. (2011). Does memory modification threaten our authenticity? *Neuroethics, 4,* 235–249.

Farah, M. J., Illes, J., Cook-Deegan, R., Gardner, H., Kandel, E., King, P., et al. (2004). Neurocognitive enhancement: what can we do and what should we do? *Nature Reviews, 5,* 421–425.

Fenton, A., Meynell, L., & Baylis, F. (2009). Ethical challenges and interpretive difficulties with non-clinical applications of pediatric FMRI. *American Journal of Bioethics, 9*(1), 3–13.

Fukuyama, F. (2002). *Our posthuman future.* New York: Farrar, Straus & Giroux.

Glannon, W. (2007). *Bioethics and the brain.* New York: Oxford University Press.

Grady, S., Aeschbach, D., Wright, K. P., Jr., & Czeisler, C. A. (2010). Effect of modafinil on impairments in neurobehavioral performance and learning associated with extended wakefulness and circadian misalignment. *Neuropsychopharmacology, 35,* 1910–1920.

Greene, J. D. (2008). The secret joke of Kant's soul. In W. Sinnott-Armstrong (Ed.), *Moral psychology* (Vol. 3, pp. 35–81). Cambridge, MA: MIT Press.

Greene, J. D., & Cohen, J. (2004). For the law, neuroscience changes nothing and everything. *Philosophical Transactions of the Royal Society, 359*(1451), 1775–1785.

Greene, J. D., Somerville, R. B., Nystrom, L. E., Darley, J. M., & Cohen, J. D. (2001). An fMRI investigation of emotional engagement in moral judgment. *Science, 293*(5537), 2105–2108.

Henry, M., Fishman, J. R., & Younger, S. J. (2007). Propranolol and the prevention of post-traumatic stress disorder: Is it wrong to erase the "sting" of bad memories? *The American Journal of Bioethics, 7*(9), 12–20.

Kaposy, C. (2010). The supposed obligation to change one's beliefs about ethics because of discoveries in neuroscience. *American Journal of Bioethics—Neuroscience, 1*(4), 23–30.

Kass, L. R. (2003). *Beyond therapy: Biotechnology and the pursuit of human improvement.* Washington, DC: The President's Council on Bioethics.

Kay, K. N., Prenger, T., Rayan, J., & Gallant, J. L. (2008). Identifying natural images from brain activity. *Nature, 452,* 352–355.

King, D. J. (1992). Benzodiazapines, amnesia and sedation: Theoretical and clinical issues and controversies. *Human Psychopharmacology: Clinical and Experimental, 7,* 79–87.

Kolber, A. J. (2007). Pain detection and the privacy of subjective experience. *American Journal of Law & Medicine, 33,* 433–456.

Lewis, D. (1972). Psychophysical and theoretical identifications. *Australasian Journal of Philosophy, 50,* 249–58.

Liao, M. S., & Sandberg, A. (2008). The normativity of memory modification. *Neuroethics, 1,* 85–99.

Libet, B. (1999). Do we have free will? *Journal of Consciousness Studies, 6*(8–9), 47–57.

Machado, C., Estevez, M., Rodriguez, R., Pérez-Nellar, J., Chinchilla, M., DeFina, P., et al. (2014). Zolpidem arousing effect in persistent vegetative state patients: autonomic, EEG and behavioral assessment. *Current Pharmaceutical Design, 20,* 4185–4202.

Meegan, D. V. (2008). Neuroimaging techniques for memory detection: Scientific, ethical, and legal issues. *American Journal of Bioethics, 8*(1), 9–20.

Mehta, M. A., Owen, A. M., Sahakian, B. J., Mavaddat, N., Pickard, J. D., & Robbins, T. W. (2000). Methylphenidate enhances working memory by modulating discrete frontal and parietal lobe regions in the human brain. *Journal of Neuroscience, 20*(6), RC65.

Monti, M. M., Vanhaudenhuse, A., Boly, M., Pickard, J. D., Tshibanda, L., et al. (2010). Willful modulation of brain activity in disorders of consciousness. *The New England Journal of Medicine, 362*(7), 579–589.

Monti, M. M., Pickard, J. D., & Owen, A. M. (2013). Visual cognition in disorders of consciousness: From V1 to top-down attention. *Human Brain Mapping, 34*, 1245–1253.

Owen, A. M., Coleman, M. R., Boly, M., Davis, M.H., Laureys, S., & Pickard, J.D. (2006). Detecting awareness in the vegetative state. *Science, 313*, 1402.

Owen, A. M., & Coleman, M. R. (2008). Functional neuroimaging of the vegetative state. *Nature Reviews. Neuroscience, 9*, 235–243.

Pitman, R. K., Sanders, K. M., Zusman, R. M., Healy, A. R., Cheema, F., Lasko, N. B., et al. (2002). Pilot study of secondary prevention of postraumatic stress disorder with propranolol. *Biological Psychiatry, 51*, 189–192.

Richmond, S. (2012). Brain imaging and the transparency scenario. In S. Richmond, G. Rees, & S. J. L. Edwards (Eds.), *I know what you are thinking: Brain imaging and mental privacy* (pp. 185–203). Oxford: Oxford University Press.

Roskies, A. (2002). Neuroethics for the new millennium. *Neuron, 35*(1), 21–23.

Sandel, M. (2004). The case against perfection. *The Atlantic, 293*(3), 51–62.

Serrano, E., & Pastalkova, P. (2006). Storage of spatial information by the maintenance mechanism of LTP. *Science, 313*, 1141–1144.

Shewmon, D. A. (1998). Chronic "brain death": meta-analysis and conceptual consequences. *Neurology, 51*, 1538–1545.

Shewmon, D. A. (2010). Constructing the death elephant: A synthetic paradigm shift for definition, criteria, and tests for death. *Journal of Medicine and Philosophy, 35*, 256–298.

Squire, E. D., & Caine, L. R. (1980). Qualitative analysis of scoplamine-induced amnesia. *Psychopharmacology, 74*(1), 74–80.

Sugden, C., Housden, C. R., Aggarwal, R., Sahakian, B. J., & Darzi, A. (2012). Effect of pharmacological enhancement on the cognitive and clinical psychomotor performance of sleep-deprived doctors. *Annals of Surgery, 255*(2), 222–227.

Synofzik, M. (2009). Ethically justified, clinically applicable criteria for physician decision-making in psychopharmacological enhancement. *Neuroethics, 2*, 89–102.

Tye, L. D., Cone, J. J., Hekkelman, E. F., Janak, P. H., & Bonci, A. (2010). Methylphenidate facilitates learning-induced amygdala plasticity. *Nature Neuroscience, 13*, 475–81.

Veatch, R. M. (2005). The death of whole-brain death: The plague of the disaggregators, somaticists, and mentalists. *Journal of Medicine and Philosophy, 30*, 353–378.

Wasserman, D. (2004). Making memory lose its sting. *Philosophy & Public Policy Quarterly, 24*, 12–18.

Weisberg, D. S., Keil, F. C., Goodstein, J., Rawson, E., & Gray, J. R. (2008). The seductive allure of neuroscience explanations. *Journal of Cognitive Neuroscience, 20*(3), 470–477.

Chapter 2
Rethinking Commonsense Conceptual Frameworks

Abstract Debates about the ethical implications of advancements in neuroscience often include estimates of how such developments will affect commonsense morality. These predictions rely on a putative clash between commonsense morality and neuroscientific discoveries. In this chapter, I argue that commonsense morality is an empirically evaluable theory, which can be circumscribed in the same way as commonsense psychology—using Lewis's method of collecting quotidian platitudes. I maintain, however, that if one were to utilize this method of collecting platitudes about morality, such a collection will represent only current commonsense morality. Commonsense morality specific to a particular time and cultural context cannot support unqualified claims that commonsense moral concepts as such are incompatible with scientific discoveries that pertain to the moral domain. Similarly, because general arguments about the character of commonsense concepts cannot be buttressed using these limited samples, commonsense moral concepts should not be used to set immutable boundaries for the development of new theories and conceptual frameworks.

2.1 Introduction

Debates about the ethical implications of advancements in neuroscience often include estimates of how such developments will affect commonsense morality. These predictions rely on a putative clash between commonsense morality and neuroscientific discoveries, and recapitulate debates already held in philosophy of mind about the seeming discontinuity between folk-psychological views about the mind and scientific approaches to psychology. In both debates, commonsense conceptual frameworks are characterized as intransigent, which motivates either an argument for the elimination of those concepts or an argument that conceptual change must satisfy common sense. I argue for the view that the utilization of a commonsense framework requires the endorsement of an empirically testable theory. I argue further that commonsense morality has absorbed the influences of scientific discovery and has changed to accommodate them.

To support my view, I utilize arguments presented by eliminativists in philosophy of mind, who have argued that our commonsense psychology constitutes an empirically testable theory. I apply this claim to commonsense morality in order to

N. Gligorov, *Neuroethics and the Scientific Revision of Common Sense*,
Studies in Brain and Mind 11, DOI 10.1007/978-94-024-0965-9_2

show that it too can be characterized as an empirically testable theory. This chapter is divided into three sections. In Sect. 2.2 of this chapter, I present the argument that folk psychology (FP), or commonsense psychology, is an empirically testable theory that posits mental states, such as thoughts, beliefs, and sensations, to predict and explain behavior. I then argue for how a view that defines mental states as theoretical posits can also account for them as experiences—in effect showing how a theory can become a conceptual framework utilized daily to interpret and predict behavior. Despite endorsing aspects of the eliminativists' view, I argue against the conclusion that commonsense psychology needs to be replaced to give way to a scientific account of human psychology. I argue that FP incorporated scientific facts about human psychology and that because of that, the incompatibility between commonsense and scientific psychological frameworks required to support the call for the elimination of commonsense concepts is not there. I argue similar conclusions apply to commonsense morality, especially as some moral concepts often designate purportedly human psychological abilities, such as free will.

In Sect. 2.3, I show how David Lewis's functionalist approach (1972) to psychophysical reduction can be used to circumscribe the boundaries of commonsense psychology, and, as I argue, can be used to establish the scope of commonsense morality as well. Lewis's method requires the collection of commonsense platitudes used in everyday parlance to predict and explain human behavior. The collection of those platitudes can in turn be taken to define mental terms implicitly and ultimately to specify a theory with covering laws. Although functionalists sometimes draw a distinction between commonsense and scientific platitudes (Block 1991, ed.), I show that such a distinction is not supportable. A compilation of our current commonsense platitudes is likely to include many scientific facts and the current commonsense theory derived from them is likely to be continuous with scientific approaches to human psychology, which preempts the need for the elimination of relevant commonsense concepts. In effect, I argue scientific facts have been and will continue to be incorporated into our everyday parlance, and therefore into our commonsense.

Because commonsense is influenced by a variety of different sources and changes over time with those influences, I argue, commonsense theories even when characterized using Lewis's method, provided only a culturally and temporally limited sample of those frameworks that should not be used to make general claims about the nature of commonsense concepts as such. Thus, I disagree with arguments that commonsense conceptual frameworks should be used to limit scientific or philosophical inquiry in the domains of psychology and morality. Commonsense should not be ignored, but it should not be accorded privileged status either.

2.2 Commonsense Psychology as and Empirically Testable Theory

In this section, I will focus on a position in philosophy of mind called eliminative materialism (EM). Although I do not endorse eliminativism for reasons described latter in this section, I accept several aspects of this view, including the argument that commonsense psychology (or folk psychology (FP)) is best characterized as an empirically testable theory. I present this way of describing FP because I wish to apply it to commonsense morality and its corollary concepts. Although commonsense morality is often mentioned in the neuroethics literature, it is rarely characterized in sufficient detail to enable a determination of the true relationship between putatively commonsense concepts and their scientific counterparts.

Several authors have been proponents of eliminative materialism. Paul Fayerabend (1962) and Richard Rorty (1979) endorsed something akin to an eliminativist position. Both authors argued that our everyday parlance featuring references to mental states could be replaced, without loss, by a discourse that features none of those references. Contemporary proponents of eliminative materialism (EM) include Stephen Stich in his 1983 text *From Folk Psychology to Cognitive Science: the Case against Belief*, and Paul and Patricia Churchland. EM is a type of materialism (or physicalism),[1] which is the view that there is only one physical substance that underlies all phenomena in the world. Physicalism or materialism can be contrasted with dualism, which is the view that there are two substances in the world, the physical substance and the mental substance, i.e., the substance that realizes our mental states. The mind and body problem designates the set of difficulties that result from attempting to apply a physicalist approach to account for the nature of mental states, which have often been characterized as having subjective properties not amenable to scientific explanation. Eliminative materialism is an attempt to resolve the mind and body problem through the elimination of mental states, such as thoughts, beliefs, and sensations.

In my description of EM, I will focus mostly on the position espoused by Paul Churchland because he has remained a proponent of this position and because he has developed the argument I wish to utilize, which is that FP is an empirically testable theory. Churchland's version of EM has two premises (Churchland 1992, pp. 2–8). The first premise is that folk psychology (FP) is an empirical theory. FP designates the commonsense view of human psychology implicit in our everyday reports of mental states such as beliefs, emotions, sensations, and attribution of those states to other people. "These are generalizations that are "common knowledge" among ordinary folk. Almost everyone assents to them, and almost everyone knows that almost everyone else assents to them" (Stich 1996, p. 127). Eliminativists argue that FP is an empirical theory because it posits mental states to explain and

---

[1] Although it is possible to establish a difference between materialism and physicalism, this difference is not relevant to my argument, and I will use the two terms interchangeably. For more on the differences between these terms, see Stoljar (2009).

predict overt behavior. In addition, eliminativists argue that FP includes law-like generalizations for the explanation and prediction of behavior (Churchland 1992, pp. 4–5).

The second premise of EM is that FP is an inadequate empirical theory. According to Churchland, there are three reasons that FP is inadequate. First, it does not compare favorably with scientific endeavors to explain human psychology. For example, it fails to explain phenomena Churchland consider to be within the domain of FP; it fails to provide explanations for mental illness, creative imagination, individual differences in intelligence, etc. (Churchland 1992, p. 7). A further problem with FP, according to Churchland, is that it never changes, and one should take this to be evidence that it is not a good theory. A look at the history of FP reveals it to be a, "…(S)tory…of retreat, infertility, and decadence" (Churchland 1992, p. 7). Churchland says the FP of the ancient Greeks and current FP is the same theory.

The second stream of challenges to FP focuses on its purported commitment to the sentence-like structure of propositional attitudes such as beliefs. It is unlikely that beliefs have a linguistic structure, because it is possible to attribute mental states akin to beliefs to infants and animals, neither of which are linguistically competent. Furthermore, research into the neural structures that underlie the organization and processing of perceptual information reveals that such processes accomplish a great variety of complex tasks, some of which show complexity far in excess of natural language (Churchland 1986, p. 396). Natural languages, it turns out, exploit only a very elementary portion of the available machinery, the bulk of which serves far more complex activities beyond the ken of the propositional conceptions of FP (Churchland 1992, p. 19).

The third and last type of attack mounted against FP is that it is a view that is committed to the Cartesian description of mental states, which is that mental states are necessarily conscious and that introspection provides veridical access to properties of mental states. Yet, according to Churchland, there is mounting scientific evidence that the Cartesian characterization of mental states is not accurate. First, there is evidence from masked priming effects that stimuli not consciously perceived nonetheless influence behavior (Merikle and Daneman 2000; Dehaene 2014). Second, there is evidence from blind-sight patients. Blind-sight patients have an intact visual system, but have damage in the occipital area of the brain, the area of the brain responsible for vision. Blind-sight patients do not report having visual experiences, but in some context behave as sighted individuals. For example, when asked to walk down a corridor, they are capable of avoiding obstacles in their path even if they cannot report consciously perceiving the objects in their way (Kolb and Whishaw 1980, p. 254). Third, experiments show that people propose erroneous verbal explanations of their own behavior, which challenges the claim that introspection is the best way to determine the real properties of mental states (Nisbett and Wilson 1997).

The conclusion of EM is that the entities posited by FP, such as thoughts, beliefs, and sensations, are illusory because FP is a false theory. Churchland's argument goes even further, proposing that we should revise FP in a way that replaces mental categories with neuroscientific concepts. Instead of invoking mental states to explain and predict behavior, we should use brain states.[2]

It might seem that the conclusion of EM merely denies what is obvious. Everyday parlance is rife with references to mental states because people experience mental states. John Searle (1992), for example, voices this concern when he argues that mental states are not posited entities, they are experiences. But eliminativists do have a way of accounting for both the experiential features of mental states and their nature as posits of a theory. Churchland's view that folk psychology is a theory is based on an account presented by Wilfrid Sellars in *Empiricism and Philosophy of Mind* (ed. 1997). Sellars, using the Myth of Jones, describes a process by which our hypothetical ancestors learned folk psychology. In this myth, Jones develops a theory that posits inner states, i.e., mental states, to explain overt behavior. Inner states are theoretical posits because they are not entirely reducible to overt behavior. Jones finds that positing those theoretical entities helps him to predict overt behavior successfully. Thus, Jones concludes, overt behavior is the culmination of a process that begins with "inner speech" (Sellars ed. 1997, p. 103).

Jones teaches his compatriots his new theory in the following manner: "Jones brings this about, roughly, by applauding utterances by Dick of 'I am thinking that p" when the behavioral evidence strongly supports the theoretical statement 'Dick is thinking that p;' and by frowning on the utterances of 'I am thinking that p,' when the evidence does not support this theoretical statement. What began as a language with purely theoretical use has gained a reporting role" (Sellars ed. 1997, p. 107). When Jones teaches his compatriots commonsense psychology, he imparts a conceptual framework that enables them not only to talk about their inner states, but to individuate them as mental states. Based on this view, the mere occurrence or experience of an inner state is not sufficient to produce first-person reports of mental states as such. Reports of inner states are made possible by our endorsement of a theory that posits mental states and the utilization of a resultant conceptual framework in everyday life. The endorsement of Jones's conceptual framework that features mental states enables us to individuate mental states as such and produce first-person reports of being in those states. Thus, the production of contemporaneous first-person reports of the form 'I think that p' requires the endorsement of a theory that posits entities such as thoughts. Jones's theoretical posits both enable first-person reports and influence their content, such that a change in theory would result in changed first-person reports. I detail Sellars's argument here because I will use it later in the book to show both how it can be used to argue that folk morality is a theory and how folk morality, when endorsed, can generate first person reports.

Sellars's view allows Churchland to respond to Searle's criticism and to explain how endorsement of a theory can influence our experiences. We experience mental states as we do because we have endorsed a conceptual framework that individuates inner states as mental states. Moreover, we are practiced at reporting mental states

---

[2]Eliminativist proposals have resurfaced in neuroethics, most notably about concepts of free will and moral responsibility. For a representative view, see: Greene and Cohen 2004.

to such a degree that we no longer consciously experience the inferential process followed by Jones when he first devises the theory. And if it is indeed the case that the way we experience our inner states is the result of learning to utilize a conceptual framework, such as FP, there is no principled reason for why we could not just learn another theory instead. Based on this view, a reconceptualization of our quotidian psychological framework is possible, including a change that would entirely supplant our daily mentions of mental states. This is because, generally speaking, the individuation and categorization of any phenomenon is the result of a process of acquiring habits of responses to stimuli in a variety of different circumstances (Sellars ed. 1997, p. 148). In effect, Sellars's view is that to have and use any one concept requires the endorsement of at least a minimal background conceptual framework. It is the adoption of a background theory that allows us to individuate experiences and other phenomena *as* belonging to a particular concept or conceptual category.[3] It is this particular aspect of Sellars's view that allows for the implication that conceptual change can result in changed experiences as well as changes in individuation and categorization of any number of phenomena. I adopt this aspect of Sellars's argument, and I apply it to assess claims throughout the book, especially in Chaps. 3, 7, and 8.

To refute EM, one could challenge EM's first premise that FP is a theory, in the same manner as Searle (1992). Or one could countenance the first premise of EM but then disagree with the second premise, which is that FP is an inadequate empirical theory. For example, one could argue that the scope of FP is not clearly demarcated and that is not always clear which phenomena are within its domain. Churchland countenances that we use words and conceptual categories such as memory, learning, and perception in our everyday parlance, but he faults FP for not accounting for phenomena that might be more accurately characterized as being within the purview of cognitive science or neuroscience. Churchland (1992) argues that the failure of FP includes the utter ignorance of the nature of sleep, the inability to account for the neural mechanisms that allow humans to catch objects in flight, the perceptual processes that lead to perceptual illusions, the "miracle of memory," and the ability humans to retrieve learned information (p. 7). Finally, one can agree with both of the premises of EM, but reject as the conclusion that FP can be eliminated. For example, some have compared FP to language, arguing that some elements of FP are innate and cannot be replaced (Carruthers 1996; Fodor 1975). I endorse the view that FP is an empirical theory and that it could be eliminated in principle. I maintain however that elimination is not required to give way to a more scientifically grounded quotidian psychology. In what follows, I argue for that view.

---

[3] For example, one does not need a conceptual background to experience a noxious sensation, but one does need at least a minimal conceptual framework to categorize this sensation as pain, i.e., to think (or report) "I am in pain." Similarly, in order to categorize an individual as having free will or as being dead, one must have at least rudimentary conceptual framework that features the concepts of 'free will' and 'death.' The particular characterization of individual concepts, as well as the properties of the corollary phenomena, is determined by the role those play in the conceptual framework and can change as the framework changes.

2.2.1 The Purported Incompatibility between Commonsense and Scientific Conceptual Frameworks

I accepted the first premise of eliminativism that folk psychology constitutes an empirically evaluable theory because it predicts and explain human behavior by positing entities such as thoughts and sensations, and that our quotidian utilization of such explanations has a law-like structure. Assuming this characterization of FP, we can conclude that all claims about human psychology derive from the commitment to some empirically testable theory. The battle between commonsense psychology and neuroscience can be recast as the battle between two scientific theories. The crux of the eliminativist argument, then, is just the prediction that neuroscience will prove to be a more superior theory than FP, and will replace it as the preferred conceptual framework in everyday life.

There are two different scenarios for how neuroscience or some other version of scientific psychology could come to replace FP. I present both of the ways to show what would be required to generate incompatibility between scientific and commonsense conceptual frameworks, and to then argue that such incompatibility does not arise. Imagine we endorse a theory much like our current commonsense psychology that predicts and explains human behavior by utilizing concepts, such as beliefs, emotions, and sensations. Imagine further that a different and new theory much like current neuroscience can explain and predict human behavior by positing physical entities such as brain states. If the two theories are compatible, the new theory could reduce the old theory.

On the model for intertheoretic reduction espoused by Nagel (1961), a new and more comprehensive theory reduces the old theory just in case the new theory, plus correspondence rules, entails the old theory.[4] Under this type of reduction, correspondence rules or bridge laws are introduced to establish identities between the entities in the old and new theory. For example, in case commonsense psychology is reduced to neuroscience, bridge laws would connect mental states posited by FP to brain states or processes identified by neuroscience.

The concepts of the old theory—here, commonsense psychology—would be preserved. A successful intertheoretic reduction of commonsense psychology to neuroscience would vindicate the entities endorsed by commonsense psychology because it would provide additional proof that those entities have a physical

---

[4] "Difficulties with this view begin with the observation that most reduced theories turn out to be, strictly speaking and in a variety of respects, *false*" (Churchland 1992, p. 48). Based on Nagel's view, from the new theory plus "bridge laws," one can deduce the old theory. But if one has an identity between old and new entities, one can get from the falsity of old entities to the falsity of the new entities. "If reduction is deduction, modus tollens would thus require that the premises of the new reducing theory be somehow false as well, in contradiction to their assumed truth" (Churchland 1992, p. 48). According to Churchland, in most cases the problem can be solved by adding a counterfactual boundary condition to the reducing theory. Doing this would confine the falsity in the premises of the reducing theory will be confined to these conditions.

instantiation in the brain. Reduction would not require the elimination of folk-psychological concepts. It would, however, entail some revision of the original conceptual framework. For example, since reduction could make it true that mental states are localized in the brain and are physical, our concepts of mental states would have to be revised to include that fact.

Intertheoretic reduction does not succeed between theories where the reduced theory is radically false, because the entities posited by it are illusory and must be rejected (Churchland 1992, p. 48). Since the ontologies of the reducing and reduced theories are connected by bridge laws, the status of these laws would be put into question if the reduced theory has an ontology that is "illusory or uninstantiated" (Churchland 1992, p. 48).

The distinction between reduction and elimination is not clear cut, however. Consider the following example: Tables and chairs are sometimes characterized as entities featured in our folk physics. Folk physics can be construed as a theory that attempts to predict and explain the behavior of middle-sized objects, such as tables and chairs. Suppose further that folk physics can be successfully reduced to physics proper. Physics does not feature entities such as tables and chairs; it posits elementary particles used to explain and predict the behavior of middle-sized objects, i.e., objects observable by the naked eye. According to some eliminativists, such as Quine (1969) and Fayerabend (1962), successful reduction entails elimination. The only entities that are real after the reduction of folk physics to physics are elementary particles, and we could in principle completely omit, without explanatory loss, the entities of folk physics. Given that even the reality of reduced entities, such as tables and chairs, can be questioned, it is difficult to distinguish them from illusory entities, such as phlogiston and witches (Lycan and Papas 1972).

Eliminativists ought to be able to draw a distinction between entities that are false because they do not exist and those that are false because things have been discovered that redefine their properties. Rorty (1979) maintains that there is not an empirical way of telling the difference between successful reduction and elimination. He argues that either you are talking about Xs but practically everything you say about them is false, or since practically nothing you say is true of Xs, you cannot be speaking about Xs (Rorty 1979, p. 80).

In Rorty's first instance, one is opting for reduction, whereby the terms of the reduced theory are preserved. For example, if neuroscience reduces psychology and mental states are identified with brain states, mental terms would continue to be used, but the way we speak of mental states would be mostly false. After reduction, all that would remain would be something akin to a nominal use of mental terms, because our real inner states would be brain states with properties unlike the properties of mental states. If, instead, one opts for elimination, the terms of the reduced theory are just dropped from our vocabulary, because the reducing theory, here neuroscience, failed to validate any of the properties often attributed to psychological states. Rorty claims that there is nothing empirical that could help us choose between the two options.

It is in part for this reason that Stich (1996) argues that the way we abandon some categories, such as witches and phlogiston, and keep others, such as atoms reducible to more elementary particles, is based on political and social factors and not on whether a theory that features those categories was eliminated or reduced. For example, that phlogiston was eliminated in favor of oxygen could be attributed to Lavoisier's wanting to be thought of as an innovator by introducing a new term instead of revising an old one (Stich 1996, p. 68).

Even if we set aside the difficulty of distinguishing between elimination and successful reduction, I maintain that an incompatibility between commonsense views about the nature of mental states and a neuroscientific explanation of them required for eliminativism is unlikely. Arguments for elimination of our commonsense psychology rest on a type of essentialism for folk-psychological concepts. In order to create an incompatibility between FP and neuroscience, one needs to support the claim that we can specify a steady endorsement of a particular commonsense view about the nature of mental states that is unlikely to change in the future.

There is instead reason to believe that commonsense psychology has changed over time. Current FP includes statements about human psychology that identify the locus of human psychology in the brain and attribute changes in behavior to processes in the brain. For example, phrases referring to changes in the chemicals in the brain are often used in everyday parlance to explain a variety of phenomena. We speak of depression being caused by imbalances in the chemicals in the brain. Runners sometimes refer to "runner's high" and explain it by invoking changes in neurotransmitters in the brain. In some cases specific neurotransmitters, such as serotonin, dopamine, and oxytocin are invoked in everyday explanations of distress or happiness, for example.

Brain damage is widely thought to be associated with changes in behavior, cognitive abilities, and moods, and that association is invoked in everyday parlance when discussing individuals with neurologic diseases such as Alzheimer's. Much of this is likely due to the popularizing of neuroscience. One could venture an argument that reliance on scientific facts and explanations of human behavior that are in some way derived from science are in fact becoming incorporated into our commonsense psychology in the way Churchland encouraged and predicted.

Yet Churchland argues the FP is static and has not changed since the time of the ancient Greeks. But contrast current folk psychology, especially as it pertains to the localization of psychological states, with Aristotle's argument that sense perception in sanguineous animals, which includes humans, originates in the heart, not the brain. He even maintains that emotions are caused by movements in the blood; for example, anger is "the boiling of the blood and hot stuff around the heart" (Aristotle 1987, ed. 163). Although in some cases one still refers to a broken heart to describe how jilted lovers feel, or describes anger as the boiling of blood, invocations of the functions of the heart as they relate to emotions in everyday psychological explanations is much of the time just a figure of speech.

Churchland also argues that FP is committed to the view the mental states are conscious states, which is a false assumption as there is evidence that thoughts or sensations can affect behavior without being conscious.[5] Contemporary FP, however, includes attribution of unconscious states. For example, in every day parlance, to explain unexpected utterances, we sometimes invoke Freudian slips. We also countenance the possibility that individuals are affected by unconscious biases, which might cause them to act as prejudiced about race, gender, age, etc. That we refer to Freudian slips or unconscious biases substantiates the claim that advances and theoretical approaches in scientific psychology do become incorporated into our everyday parlance, and that FP explains human behavior in terms of unconscious mental states.

Changes in ontological commitments of folk psychology are an obstacle for the argument that the categories of common sense are illusory because it undermines the claim that commonsense psychology can be characterized and circumscribed correctly. Any characterization of common sense ought to be limited to a particular time and context, as commonsense concepts depend on culture (Stich 1998).[6] One could perhaps argue that the FP of the Greeks could be eliminated in favor of neuroscience because its categories are illusory, but the same claim might not as easily fit current FP. The argument that FP has an illusory ontology becomes further complicated by the argument that science does influence FP, and that FP has changed as a result of discoveries in neuroscience. If current FP includes the commitment to brain processes and uses those to account for human behavior, then its ontology is not illusory, because some of it is shared with neuroscience.

2.3   Characterizing Commonsense Concepts

To recapitulate my argument thus far, I used Churchland's argument to show that FP is an empirically testable theory. I then showed how, based on Sellars's argument, a conceptual framework, like FP, could give rise to experiences of the sort that result in first-person reports. This was done to support the argument that mental states, although they are experienced, can also be rightly characterized as theoretical posits of a commonsense psychology. In addition, I accepted Sellars's argument that a minimal background theory or conceptual framework is required for the individuation and categorization of experiences as well as other phenomena. I will now move to describe how the domain of a commonsense psychology could be circumscribed and how mental states are defined by their role in that commonsense theory. I will then apply this manner of circumscribing commonsense theories to commonsense morality. Finally, I will show how this way of demarcating commonsense theories supports my argument against the incompatibility between commonsense and

---

[5] Examples of this were presented earlier in this section.

[6] For more on the fact that commonsense categories can depend on culture, see Stich 1998. Also see, Dennett (1987, p. 54), who argues that folk psychology varies, not just across cultures, but even within countries, states, or neighborhoods.

scientific theories, which generate the argument for the elimination of common-sense concepts in psychology and morality.

To do all this, I will reprise David Lewis's approach to functionalism and his proposal for psychofunctional reduction, which relies on the implicit functional definition of mental terms. David Lewis proposes that by collecting everyday psychological platitudes, we can arrive at an implicit functional definition of mental terms (Lewis 1972, pp. 249–250). "Collect all the platitudes you can think of regarding the causal relations of mental states, sensory stimuli, and motor responses. ... Add also all the platitudes to the effect that one mental state falls under another—'toothache is a kind of pain,' and the like. ...Include only platitudes which are common knowledge among us—everyone knows them, everyone knows that everyone else knows them, and so on. For the meanings of our words are common knowledge and the names of mental states derive their meaning from these platitudes" (Lewis 1972, pp. 255–256). The collection of the relevant platitudes constitutes our commonsense psychology, which defines mental terms, including beliefs, desires, and sensations. Furthermore, the collected platitudes describe the interaction between mental states and the way in which they mediate between stimuli and motor responses, which in turn specifies the laws of our folk psychology.

The theory specified by our psychological platitudes will feature two different kinds of terms, theoretical terms (T-terms for short) and observational terms (O-terms for short). Mental terms are called T-terms because they are introduced by the theory specified by the collected platitudes. The meaning of T-terms is not known prior to collecting the platitudes that comprise the theory. All the other terms referenced in the platitudes are O-terms. The meaning of O-terms is known prior to the introduction of the theory about mental terms because they have been fixed by other theories. The O-terms of the collected platitudes specify the function of the T-terms, which is how they implicitly define the T-terms.

For example, consider the following story: Dr. Franklin arrives to examine a patient. The patient is lying motionless in the bed. Dr. Franklin suspects the patient is dead. He believes that feeling the patient's pulse would help him determine whether the patient is dead. When he places his index finger at the base of the thumb on the patient's wrist, he cannot feel her pulse. Dr. Franklin doubts the accuracy of his measure and reaches for the stethoscope because he knows that it will help him determine whether the patient has a heartbeat.

In this story, the T-terms would be those referring to Dr. Franklin's mental states such as 'feels,' 'suspects,' 'believes,' 'doubts,' and 'knows.' The observational terms are most of the others, including 'stethoscope,' 'pulse,' 'heartbeat,' 'patient,' etc. The meanings of the O-terms have been defined by other theories; for instance, the meaning of the term 'heartbeat' is defined by a biological theory that specifies the role of the heart in the functioning of the human organism. The definition of the term 'stethoscope' and the usefulness of the device in determining the presence and pace of a heartbeat have been set by other theories as well. The meanings of the T-terms, i.e., Dr. Franklin's mental states, are set by the role of those in the story above. For example, that Dr. Franklin believes that the patient is dead explains why he checks her pulse. Furthermore, that he suspects a manual determination of it is not an adequate measure, and that he knows a stethoscope is more reliable, leads him to reach for the device.

Once the relevant platitudes are collected, commonsense psychology could be uniquely realized, near-realized, or not realized at all (Lewis 1972, p. 252). That is, the implicit definition of the mental terms can be true of something in particular, mostly true of something, or not true of anything in the world. Thus, the specific choice of psychological platitudes is important because they determine the overall truth of the theory. When scientific psychology and FP are contrasted, most often the comparison ends badly for commonsense psychology because its scope is restricted to unscientific characterizations of mental states. Churchland characterizes FP in this manner. Ned Block (1991 ed.) does something similar by distinguishing between functionalism and psychofunctionalism. He describes functionalists as being limited to the platitudes derived from our commonsense psychology, and classifies Lewis as a functionalist. Psychofunctionalists do not have any such restrictions. They can rely on facts about psychology derived from scientific psychology and even from neuroscience.[7]

But Lewis's method specifies only that in order to determine the boundaries of FP, we should collect platitudes about human psychology used in everyday life by lay persons. Thus, something is a commonsense platitude because many of us know it and use it in everyday life to explain and predict aspects of human behavior. Lewis's way of selecting commonsense platitudes remains vulnerable to the criticism that any selection of platitudes will not be representative of FP in general. The scope of FP will depend on the selection of a representative sample. And if it is true that historical context and cultural differences can influence how we interpret and predict human behavior, then any sampling of commonsense platitudes will have limited generalizability.

But I wish to make a further point, which is that if the way in which we determine the scope of commonsense psychology is by collecting platitudes used in everyday parlance, there is no guarantee that scientific facts will not be included in the collection. This would be true, even if one were to exclude neuroscientists or psychologists from our population sample. If scientific facts derived from scientific psychology or neuroscience are part of our everyday parlance, then they would become incorporated into the characterization of FP. This in turn would blur the line between commonsense and scientific characterization of psychology. If the criterion used to determine whether a platitude is common sense is that many of us know it and use it in everyday life, then any scientifically sourced platitudes we know and

---

[7] The distinction between functionalism and psychofunctionalism reflects the division of scientific inquiry into levels, such as psychology, neuroscience, biology, and so forth. For example, if the functional explanation is at the level of psychology, then the inputs and outputs are specified in terms of observable stimuli and behavior. If the explanation is at the level of neuroscience, the inputs and outputs could be specified in terms of neurotransmitters and their corollary effects. But I argue commonsense platitudes include facts from a variety of levels of explanation.

use in everyday life to explain and predict behavior would be considered part of our commonsense psychology.

Lewis (1972) describes mental terms or T-terms as pre-theoretical because they are introduced by the commonsense psychological theory. Hence, we could propose a distinction between commonsense platitudes and scientific ones based on that feature: commonsense platitudes are those that are pre-theoretical.

In the section on eliminative materialism, I presented Sellars's (1997) argument, which he describes using the Myth of Jones.[8] The conclusion of Sellars's argument, which I find convincing, is that the utilization of any term, and the individuation of any phenomenon or state, requires familiarity with a background conceptual framework. The source of the conceptual framework is a theory for which we provide endorsement by utilizing that conceptual framework. As Sellars argues, in order to identify and report a mental state, or ascribe it to another person, we must, at least implicitly, endorse a theory that posits mental states to explain and predict behavior. Based on this view, ascriptions of mental states and the platitudes that refer to them are not pre-theoretical. Hence, the distinction between pre-theoretical and theoretical beliefs cannot help restrict the scope of the selected platitudes to only those that are common sense. We would instead fall back on the criterion described previously, that commonsense platitudes are those we all know and use to explain and predict behavior. This in turn would not restrict platitudes to only those that feature terms such as beliefs, desires, and sensations. Perhaps once one accepts that the way in which we interpret and predict human psychology in everyday life constitutes a theory, its characterization as common sense should become obsolete. Omitting that designation, however, is not required for my argument.

A further counterargument could be that commonsense platitudes should be those that refer only to mental states, such as beliefs, desires, emotions, and sensations. One could argue that those terms might not be pre-theoretical, but they are mental, and mental states are those posited by our commonsense psychology. This line of argument, however, ignores the possibility that any of our mental terms could be reduced to a scientific theory. For example, even Churchland (2005) has argued that for some mental states such as color sensations, we can have a successful reduction to brain processes occurring in the visual cortex. In that case, the term 'sensation' would refer to a particular brain process, and could be properly thought of as defined by both our folk psychology and by neuroscience. To use Lewis's parlance, 'sensation' could be thought of as a T-term in both neuroscience and folk psychology. We would then have an ambiguity about whether sensations should be included into commonsense psychology regardless of whether we know and use the term in everyday parlance.

Similarly, we could conceive of a state of affairs where neuroscientists identify a particular neural correlate for other purportedly mental states, say, an emotion, such as fear. If that were that case then 'fear' would no longer be just a mental term, but a term that properly refers to a particular process in a part of the brain, such as the amygdala. This in turn would raise a similar ambiguity about whether we should

---

[8] See this chapter, Sect. 2.2, pp. 7–9.

include platitudes that contain the term 'fear' into a collection of commonsense platitudes. One can see that if the requirement for inclusion in commonsense platitudes is that mental terms be defined only by one theory, i.e., folk psychology, the scope of the platitudes would be too restricted and might exclude statements that should be properly counted among those of our folk psychology.

Selecting only platitudes that feature mental terms as T-terms would also exclude any statements that introduce T-terms that are not included in our original list of mental terms. Such a restrictive criterion would exclude statements where brain processes are introduced to explain human behavior by laypersons in everyday life. For example, "Francesca has a chemical imbalance in her brain and this is why she has trouble getting out of bed to play with her kids" or "The adrenaline spike from speaking in public has made me all jittery." In these statements it is the terms 'chemical imbalance' and 'adrenaline' that play the role previously held by mental terms. Given that, it would seem unjustified to exclude them from the collection of platitudes.

Perhaps one could argue that my cited examples are not platitudes; they are not known and utilized by most people. Because I have argued that FP is variable and can change with culture and context, it is probably never true that any platitudes are used by most people. Rather, the collection of platitudes constitutes the FP of a particular population, and I am venturing an empirical claim that in some populations, brain processes are becoming incorporated into everyday parlance. But even if commonsense psychology does not feature brain states as T-terms, it is enough for it to include them as O-terms in order for my argument to go through. Lewis argues that because T-terms, i.e., mental states, are implicitly defined by the causal role they play, the T-terms are, in principle, eliminable (Lewis 1972, p. 254). If we know the meaning of the O-terms and we use them to specify the causal role of the T-terms, it is all we need in order to determine whether a theory we have specified is realized and therefore true. Thus, what matters for the accuracy of a theory is not the use of particular T-terms, whether mental or neuroscientific, but the causal role occupied by those in the theory. If enough O-terms defined by neuroscience are incorporated into our commonsense psychology, then those will redefine the causal role of our mental terms. This in turn will increase the compatibility between neuroscience and commonsense psychology.

I think the argument I have put forth for mental states applies equally well to moral concepts. The use of moral concepts in everyday parlance, the way in which they are used to explain and predict human behavior, constitutes a theory in a way similar to that of folk psychology. Collecting platitudes that feature moral concepts as T-terms could enable us to arrive at an implicit functional definition of those terms and approximate the scope of folk morality. Folk concepts defined in this way could then be contrasted with scientific characterizations of moral concepts. As I have argued in Sect. 2.2, one may justify the elimination of a particular concept only if it is illusory or radically false. But whether a concept is radically false is determined relationally. A folk moral concept is radically false if it is defined very differently from a scientifically or philosophically defined concept.

The incompatibility between our commonsense conceptions of moral concepts and the scientific conception of those is artificially created. Commonsense notions, as I have argued, are not static, and they are at times influenced by scientific discovery. A solution that calls for the elimination of our commonsense conceptual framework is a view committed to the assumption that there is a static commonsense view that ascribes essential properties to our moral concepts. Similarly, the claims that moral concepts cannot change, or be replaced, rely on certain features being necessary for the maintenance of the concept.

To illustrate how ascribing essential properties could lead to elimination, let us use Rorty's (1979) argument that mental states are marked by incorrigibility.[9] Rorty argues that a state that is incorrigible is a mental state and nothing else is. In other words, incorrigibility is both necessary and sufficient for some state to be a mental state. If one can prove that physicalism is true and that because of that no physical states are incorrigible, as Rorty argues, then one must maintain that there are not any mental states because there are no physical states that are incorrigible. Rorty argues that the characterization of mental states as incorrigible is part of our everyday parlance, and the arguments I have presented against the claim that FP is static apply similarly to the claim that everyday parlance about mental states is static.[10] In any case, ascribing necessary and sufficient properties to mental entities makes it easier to argue for elimination.

Essentialism, however, is not supported based on a functionalist view like Lewis's, because the properties of any states or entities are defined relationally. So, if the causal role changes, the properties we ascribe to a particular phenomenon change as well. Thus, changes in a particular theory over time will affect how the conceptual categories of that theory are characterized. Given that my argument has been that common sense has been influenced by science in general, and neuroscience in particular, the current categories of common sense are likely to be more compatible with neuroscience obviating the need for elimination.[11]

There are several examples of research in neuroscience that have been interpreted as a challenge to our commonsense notions about morality as they pertain to free will, personhood, and so forth. For example, the concept of free will has been

---

[9] Incorrigibility as the mark of the mental will be covered further in Chap. 7.

[10] For more on this, see Chap. 7, especially Sect. 7.4.

[11] One could argue that without restricting commonsense conceptions to at least some required properties, it would be difficult to distinguish between slightly erroneous concepts, say "Free will is conscious willing," and completely erroneous ones, such as "Free will is the color blue." If our concepts are not construed as capturing essential properties, but acquire their properties by serving a particular functional role, then any property could be ascribed to any concept. (I wish to thank an anonymous reviewer for this comment.)

Abandoning the notion of essential properties need not entail that we relinquish means of evaluating the accuracy of our concepts; we would do that by assessing the quality of the theory in which they feature. Any theory can be evaluated by a variety of different factors including explanatory power, parsimoniousness, internal consistency, and coherence with other theories. Theories that best fit those criteria are likely to introduce the most accurate concepts, and those are not likely to be theories that radically re-describe our concepts in the manner illustrated above.

challenged by studies done by Benjamin Libet et al. (1983), which showed that unconscious preparatory brain activity precedes conscious awareness of making a choice. Also, Martha J. Farah and Andrea S. Heberlein (2007) have argued that the concept of 'person' is illusory because our ability for face recognition, detection of human bodily movements, and even our attribution of mental states to others, is automatic and innate.

If we consider free will, for example, we find a range of philosophical views that provide different formulations of that concept. For example, free will can be defined in the manner Kant does, as the ability to be moral by acting in accordance with the moral law (Kant ed. 1964). Alternatively free will can be defined according to Harry Frankfurt (1971) as the ability to have higher-order desires that can supersede our first-order desires, such as our more base desires for a variety of earthly pleasure. Or free will can be defined as Patricia Churchland (2006) does, as choices made deliberately, knowingly, and intentionally. The difficulty in formulating and agreeing on the most accurate philosophical views for just one moral concept is an indication of how difficult it is to arrive at a unitary concept of free will. And if philosophers have not been able to formulate and agree on a definition of free will, it is even less likely that an adequate and universally acceptable concept of free will can be found in our commonsense platitudes.

Without a consensus about the correct way to characterize free will, calling for the elimination of free will as such seems premature. One can of course call for the elimination of a concept defined in a particular way. So it might be easier to undermine a philosophical view of free will, for example, because philosophers aim to define and characterize free will in a specific way. Commonsense platitudes, however, do not result from a concerted effort to describe a particular phenomenon consistently.

Given that philosophy has sometimes been conceived as having the task to analyze ordinary language and perhaps produce a better understanding of our shared concepts, one could argue that the different philosophical accounts described above are all attempts to capture some shared concept of free will.[12] A philosophical treatment of free will, however, is meant to yield a depiction of the nature of free will, and analysis of commonsense concepts is not the right tool to achieve that goal.[13]

If one accepts the view that utilization of commonsense concepts requires endorsing a theory, then an exploration of our commonsense concepts might lead to a better understanding of that tacitly endorsed theory. This might be a fruitful endeavor if the goal is to characterize that theory. But there is no reason to assume that a commonsense theory is accurately capturing the nature of free will. Thus, if what one is interested in is constructing the most adequate theory of free will, one

---

[12] I wish to thank an anonymous reviewer for this suggestion.

[13] Analytic philosophy is sometimes described as the analysis of ordinary language. A thorough discussion of the proper way to conceive of philosophy is outside of the scope of my project, but I do concur with Williamson (2007, p. 21) that philosophy ought not to be conceived as a linguistic or conceptual inquiry. For a thorough consideration of why philosophy ought not to be conceived that way, see Williamson 2007.

might opt to begin by considering the commonsense concepts of free will, but if it proves inadequate, then one could opt to develop an alternative theory, which might be different from our commonsense views.

Another way of arguing that philosophical accounts should capture common concepts is to maintain that they encapsulate certain universal intuitions about a particular domain, say, morality, for example. Appeal to the use of intuitions, although ubiquitous in philosophy, is problematic.[14] There is controversy about the nature of intuitions, with many philosophers agreeing that they are beliefs without a special epistemological status (Lewis 1983; van Inwagen 1997). To illustrate how intuitions are sometimes introduced, Gopnik and Schwitzgebel (1998) provide a distinction between intuitions as hypothesis and intuitions as data, and I will apply that distinction here.

Intuitions may be used to designate a plausible initial hypothesis about a particular phenomenon, for example, that free will requires conscious decision-making. Intuitions construed as an initial hypothesis might be adopted as true, but only until there are reasons to abandon them and adopt an alternative hypothesis. If intuitions about moral concepts are thought of this way, they can be revised as more plausible hypotheses are generated. Intuitions thus characterized are not robust enough to support the argument that philosophical theories aim to capture lay moral intuitions, even in cases where those might be shared.

If intuitions are construed as data, they must be taken into account by any adequate moral theory, just like any successful empirical theory has to take into account the relevant data within its domain. For example, if it is the case that individuals have a shared intuition that free will requires conscious decision-making, then any philosophical or empirical theory will have to account for that intuition. Construing intuitions as data, however, is problematic because even those who defend intuitions as a different way of knowing admit that they can turn out to be false (Bealer 1996). Thus, an intuition that a particular phenomenon, for example, free will, has a certain property, such as conscious access, is not enough to justify that free will requires conscious access. If that is the case, one should not construe the search for the most accurate concept of free will as an attempt to capture common intuitions about that concept.

Introducing intuitions as a way of characterizing shared concepts does not support the view that philosophical theories must capture those intuitions when constructing a theory. If described as hypotheses, the appeal of intuitions trails their plausibility as a hypotheses, and permits discarding them if a more plausible explanation is formulated. Intuitions as data can be useful only if what we are studying is the existence and prevalence of certain intuitions, say, what non-philosophers think is the right way to characterize free will. If the task is to characterize the nature of free will correctly, there is no compelling reason to rest that view on common intuitions about free will, because those could turn out to be false. This is of course not

---

[14] For a comprehensive and critical assessment of the use of intuitions in philosophy, see Cappelan (2012).

an argument that our intuitions—construed as hypothesis with some initial plausibility—should not be considered.

Finally, one could maintain that all philosophical attempts to capture free will are attempts to capture the same phenomenon. Construing the philosophical endeavor in that way is not incompatible with there being a number of different theories aiming to characterize free will accurately. It is possible that there is free will, but all of our theories have failed to capture it. Or it could be that one of our theories has captured it, but we have failed to reach consensus about its adequacy in describing free will. Thus, it is still possible to argue that because there are different, sometimes contradictory, views about the nature of free will, general claims that neuroscience is showing that there is no such thing as free will is unsupported. What neuroscience can show, however, is that a particular conception of free will is false, which is a good enough outcome.

To recapitulate, Lewis's functionalism provides us with an explanation for how the theoretical terms of any theory can be defined by the functional role specified by the other terms of the theory, and in particular provides an explanation for how commonsense psychology implicitly defines mental states. The functionalism that Lewis formulates allows us to circumscribe current common sense by collecting contemporary platitudes and to characterize the entities and their properties of commonsense theories. But as our commonsense platitudes about psychology and morality change over time, the definite commitments of our commonsense psychology are indeterminate. This protects common sense from the charge that it posits entities that directly oppose a scientific psychology or neuroscience.

2.4   Conclusion

In this chapter, I present an argument for why commonsense psychology is an empirically evaluable theory. I also show how such a theory could lead to a conceptual framework that is utilized in everyday reports and ascriptions of mental states and used to predict and explain behavior. I also argue that commonsense psychology overtime has absorbed the influences of scientific theories with relevant domains and that it should be seen as continuous rather than incompatible with scientific approaches to psychology. This, I argue, undermines the need to eliminate purportedly commonsense concepts. I also argue that we can utilize Lewis's psychofunctionalism to circumscribe the boundaries of commonsense psychology and explain how mental states acquire their properties by occupying the functional roles specified by their background theory. In turn, this shows how a change in theory affects the functional role of the relevant commonsense categories or concepts and how the properties ascribed to such concepts can change as the theory changes.

I also apply this approach to commonsense morality and conclude that it too can be construed as an empirically evaluable theory, which can be circumscribed in the same way as commonsense psychology, and that the boundaries of commonsense moral concepts can be determined in similar ways as those of commonsense psychology. I argue, however, that if one were to utilize Lewis's method of collecting platitudes, such a collection will represent only current commonsense morality specific to a particular time and cultural context and thus, any such rendition of commonsense should not be used to support general arguments about the character of commonsense concepts as such. Finally, because general claims about the nature of commonsense concepts, and in particular commonsense morality, are not supportable using temporally and socially limited samples, the claims that such concepts are incompatible with scientific discoveries that pertain to moral concepts is not substantiated either. Similarly, because general arguments about the character of commonsense concepts *as such* cannot be supported using these limited samples; commonsense concepts should not be used to set immutable boundaries for the development of new theories and conceptual frameworks.

References

Ackrill, J. L. (1987, ed). *A new Aristotle reader*. Princeton: Princeton University Press.

Bealer, G. (1996). On the possibility of philosophical knowledge. *Philosophical Perspectives, Metaphysics, 10*, 1–34.

Block, N. (1978, 1991 ed). Troubles with functionalism. In D. Rosenthal (Ed.), *The nature of mind* (pp. 211–228). New York: Oxford University Press.

Cappelan, H. (2012). *Philosophy without intuitions*. Oxford: Oxford University Press.

Carruthers, P. (1996). *Language, thought and consciousness*. Cambridge: Cambridge University Press.

Churchland, P. S. (1986). *Neurophilosophy*. Cambridge, MA: MIT Press.

Churchland, P. M. (1992). *A neurocomputational perspective: The nature of mind and the structure of science*. Cambridge, MA: MIT Press.

Churchland, P. (2005). Chimerical colors: Some phenomenological predictions from cognitive neuroscience. *Philosophical Psychology, 18*(5), 527–560.

Churchland, P. S. (2006). Moral decision-making and the brain. In J. Illes (Ed.), *Neuroethics: Defining the issues in theory, practice, and policy* (pp. 3–16). New York: Oxford University Press.

Dehaene, S. (2014). *Consciousness and the brain: Deciphering how the brain codes our thoughts*. New York: Viking Penguin.

Dennett, D. (1987). *The intentional stance*. Cambridge, MA: Bradford Books.

Farah, M. J., & Heberlein, A. S. (2007). Personhood and neuroscience: Naturalizing or nihilating? *American Journal of Bioethics, 7*(1), 37–48.

Fayerabend, P. (1962). Explanation, reduction, and empiricism. In H. Feigl, M. Scriven, & G. Maxwell (Eds.), *Minnesota studies in the philosophy of science* (Vol. 3). Minneapolis: University of Minnesota Press.

Fodor, J. (1975). *The language of thought*. Cambridge, MA: Harvard University Press.

Frankfurt, H. G. (1971). Freedom of the will and the concept of a person. *The Journal of Philosophy, 68*(1), 5–20.

Gopnik, A., & Schwitzgebel, E. (1998). Whose concepts are they, anyway? The role of philosophical inuition in empirical psychology. In M. R. DePaul & W. Ramsey (Eds.), *Rethinking intuitions: The psychology of intuitions and its role in philosophical inquiry* (pp. 75–91). Lanham: Rowman & Littlefield Publishers, Inc.

Greene, J., & Cohen, J. (2004). For the law, neuroscience changes nothing and everything. *Philosophical Transactions of the Royal Society, 359*(1451), 1775–1785.

Kant, I. (1785, 1964 ed). *Groundwork of the metaphysics of morals* (H. J. Paton, Trans.). New York: Harper Torchbooks.

Kolb, B., & Whishaw, I. Q. (1980). *The fundamentals of human neuropsychology*. New York: W.H. Freeman and Company.

Lewis, D. (1972). Psychophysical and theoretical identifications. *Australasian Journal of Philosophy, 50*(3), 207–215.

Lewis, D. (1983). *Philosophical papers, volume 1*. Oxford: Oxford University Press.

Libet, B., Gleason, C. A., Wright, E. W., & Pearl, D. K. (1983). Time of conscious intention to act in relation to onset of cerebral activity (readiness potential): The unconscious initiation of a freely voluntary act. *Brain, 106*, 623–642.

Lycan, W., & Papas, G. (1972). What is eliminative materialism? *Australasian Journal of Philosophy, 50*(2), 99–105.

Merikle, P., & Daneman, M. (2000). Conscious vs. unconscious perception. In M. Gazzaniga (Ed.), *The new cognitive neurosciences* (2nd ed., pp. 1295–1304). Cambridge, MA: MIT Press.

Nagel, E. (1961). *The structure of science*. New York: Harcourt, Brace, and World.

Nisbett, R., & Wilson, T. (1997). Telling more than we know: Verbal reports on mental processes. *Psychological Review, 84*, 231–259.

Quine, W. V. O. (1969). Epistemology naturalized. In *Ontological relativity and other essays*. New York: Columbia University Press.

Rorty, R. (1979). *Philosophy and the mirror of nature*. Princeton: Princeton University Press.

Searle, J. R. (1992). *The rediscovery of the mind*. Cambridge, MA: MIT Press.

Sellars, W. (1977, 1997 ed.). *Empiricism and the philosophy of mind*. Cambridge, MA: Harvard University Press.

Stich, S. (1983). *From folk psychology to cognitive science: The case against belief*. Cambridge, MA: MIT Press.

Stich, S. (1996). *Deconstructing the mind*. Oxford: Oxford University Press.

Stich, S. (1998). Reflective equilibrium, analytic epistemology and the problem of cognitive diversity. In M. R. DePaul & W. Ramsey (Eds.), *Rethinking intuition: The psychology of intuition and its role in philosophical inquiry* (pp. 95–113). Lanham: Rowman & Littlefield Publishers, Inc.

Stoljar, D. (2009). Physicalism. In E. N. Zalta (Ed.), *The Stanford encyclopedia of philosophy*. http://plato.stanford.edu/archives/fall2009/entries/physicalism/

van Inwagen, P. (1997). Materialism and the psychological-continuity account of personal identity. In J. Tomberlin (Ed.), *Philosophical perspectives, mind, causation and world* (Vol. 11, pp. 305–319). Atascadero: Ridgeview Publ. Co.

Williamson, T. (2007). *The philosophy of philosophy*. Oxford: Blackwell Publishing.

Chapter 3
The Common Notion of Free Will

Abstract A number of studies within the domain of neuroscience have shown that conscious awareness of the decision to perform an action is preceded by unconscious activity in the brain. This in turn is taken to indicate that unconscious brain activity is the cause of action and not conscious willing. In this chapter, I assess arguments that unconscious brain activity is a threat to the common notion of free will. I dispute the idea that the common view of free will requires conscious willing. Additionally, I argue for the claim that unconscious processes play a role in the formation of conscious volitions. Based on that, I argue against the view that volition must be conscious. In this chapter, I also tackle the purported incompatibility between free will and scientific determinism. After assessing calls for the elimination of the commonsense concept of free will, I conclude that the incompatibility between the two notions rests on an unfavorable characterization of commonsense free will. I further argue that any concept of free will requires endorsement of a particular background theory and because of that I question whether any such concept can be properly characterized as common sense.

3.1 Introduction

Free will can be defined as the ability to do otherwise. This ability is thought to be particularly important for morality as it undergirds ascriptions of praise or blame. In order to hold individuals responsible for their actions, we presume that they could have acted other than they did. There have been a number of studies within the domain of neuroscience and cognitive science that have been interpreted as undermining of the assumption that humans are capable of free will. In particular, Benjamin Libet has performed a number of experiments that show that our conscious awareness of the decision to perform an action is preceded by unconscious activity in the brain, indicating that it is the unconscious brain activity that is the cause of action and not conscious willing. These results have been interpreted by Libet (1999) as showing that the commonsense concept of free will is flawed.

   Also mentioned by Libet is the purported conflict between scientific determinism and free will. The problem emerges when determinism in science is coupled with the assumption that all natural phenomena can be properly explained using scientific laws. Neuroscience has not created the problem between free will and

© Springer Science+Business Media B.V. Dordrecht 2016
N. Gligorov, *Neuroethics and the Scientific Revision of Common Sense*,
Studies in Brain and Mind 11, DOI 10.1007/978-94-024-0965-9_3

determinism, but advancement in that field has bolstered the claim that deterministic scientific laws could be used to account for human psychology, which in turn would be particularly dangerous for free will conceived of as a psychological ability.

In this chapter, I focus on both challenges to free will. The approach I take is to examine the claims that scientific progress in neuroscience undermines the commonsense concept of free will. In Sect. 3.2, I present a number of scientific findings that target the notion of free will. In particular, I describe the studies conducted by Benjamin Libet. In Sect. 3.3, I argue that those experiments do not actually challenge the commonsense concept of free will. I argue that if there is a threat to free will, it is specific to accounts that condition free will on consciousness. I then show that the commonsense notion of free will allows for ascriptions of volition even to actions that are performed without conscious willing. In Sect. 3.4, I address the purported incompatibility between the commonsense concept of free will and scientific determinism. I describe two types of positions about the commonsense concept of free will. One kind of view is held by eliminativists, who contend that the commonsense notion of free will is false and is incompatible with the truth of scientific determinism. The other type of argument is a defense of commonsense intuitions about free will. The defenders of the commonsense concept of free will maintain that it is compatible with scientific determinism and that it should be used to build an adequate theory of free will. I argue that both camps are misguided because of the assumption that the commonsense concept of free will can be circumscribed and characterized adequately enough to either prepare for elimination or to utilize as a basis of a theory.

3.2   Evidence against Free Will

Benjamin Libet defines the common notion of free will as having two elements (Libet 1999): The first element is the idea that a volitional action needs to be endogenous and free of any external control. External control can mean many different things. Often it refers to actions that are not coerced, but for Libet, an action free of external control is an action that occurs from within a person. The second element is the conscious experience of willing, or the notion that an act is free if the agent has *the feeling of wanting to do it*. In addition, this feeling needs to be the cause of the action, not just to co-occur with the decision and the action.

Libet performed a series of experiments to investigate this notion of conscious willing. In one experiment, Libet et al. (1982) asked the participants in the study to flick their wrist whenever they felt like it. The participants in the study were simultaneously monitored by an electroencephalogram (EEG) and an electromyogram (EMG). The EEG machine records electrical currents in the scalp, which are correlates of brain activity. The EMG machine detects the electrical currents in the subject's hand, caused by the actual movement of the muscles of the wrist. In the study, the movement of the wrist, and the activity recorded on the EMG was preceded by an electric charge that was recorded by the EEG—the brain became active before the wrist. This burst of electrical activity in the scalp was called the 'readiness potential' (RP). The RP in these experiments preceded the movement of the wrist by an average of 550 ms (Libet et al. 1982).

Libet et al. (1983) were also interested in measuring the conscious intention to perform the action. He called this *the first awareness of the wish to act* (W). In order to capture W, Libet and his colleagues constructed an oscilloscope clock. On the clock, a spot of light revolved around the periphery of the clock faster than the usual 60-s sweep of the second hand of the clock. The spot of light made a full circle in just 2.56 s. Each second on the oscilloscope was about 43 s of real time. The subjects were told to look at the center of the clock. For each voluntary wrist flexion, the subjects were asked to indicate where the moving spot on the clock was located when they first experienced the conscious intention to move their wrist. This procedure was intended to capture the time the subjects had experienced W (Libet et al. 1983).

In this study, as in the previous one, there was a lag between the muscular activity in the wrist and RP by about 550 ms on average. Surprisingly, there was also a lag between W and RP. The unconscious preparatory brain activity preceded the conscious intention to perform the action. This finding seems to fly in the face of the common notion, as described by Libet (1999), that conscious intention is required for voluntary action. In an attempt to accommodate the finding, Libet proposed that, although conscious willing is not the cause of the preparatory brain activity that precedes the action, once the action is activated, the will has veto power. In other words, the brain can ready us for certain kinds of actions, but the will can inhibit some actions from completing. Libet argues that the conscious veto is a control function and not just a mere awareness of the ongoing processing in the brain. Furthermore, he argues against the idea that even the conscious veto is preceded by an unconscious brain process. Libet dissociates the unconscious processes perhaps necessary for the veto from the content of the decision. He argues that, there might be some preparatory activity in the brain necessary for one to make the decision to veto or not to veto, but the actual content, to veto, for example, has to be due to the conscious will.

This last hypothesis was actually tested by Soon et al. (2008) in a study using functional magnetic resonance imaging (fMRI). FMRI capture the amount of oxygen in the blood. Oxygenated and deoxygenated blood emits different magnetic signals that are then captured by fMRI. The presumption is that active areas of the brain require more oxygenated blood, which allow for fMRI to capture brain activity as it is occurring.[1] Libet performed his experiments before this technology was available. Soon et al. (2008) had as part of their aim to replicate Libet's results using the new technology.[2] In addition, they wished to test the claim that the content of the action is not determined by unconscious brain processes, but by the conscious will.

---

[1] For a more detailed description of how fMRI records brain activity, please see Chap. 6, Sect. 6.2.
[2] For additional studies that confirm Libet's results, see Banks and Ischam (2009); Lau et al. (2007).

In the Soon et al. study, subjects were asked to fixate on the center of the computer screen and a stream of letters was presented to them. They were asked to press one of two buttons using either their left or right index fingers at any point they had the urge to do so. To capture conscious intent, Soon et al. asked the subjects to remember the letter that was on the screen when they first felt the pangs of the conscious will. Both the left and the right responses were pressed equally often and almost 89 % of subjects reported having formed a conscious intention to move in 1,000 ms before the movement (Soon et al. 2008).

Soon et al. (2008) determined that, using only the functional MR images, they could determine which action would be performed, i.e., whether the participant would press the right or the left button. This is contrary to the claim that the activity in the brain preceding the action is nonspecific preparatory motor activity because Soon et al. were able to guess the action based on brain activity alone. Moreover, they were able to predict what the subjects were going to do before they actually experienced the conscious intention to press either the left or right button. The predictive neural activity preceded the conscious decision by about 10s. Even more surprisingly, the brain activity could be used to predict the timing of the decision as early as 5 s before the action was performed. Thus, the Soon et al. study, using fMRI, confirmed Libet's results that RP precedes W, but the findings seem to undermine Libet's modification of the concept of free will as conscious veto.

There is further evidence for the dissociation of motor behavior and conscious willing. One such dissociation is illustrated by Penfield's finding that certain kinds of behavior could be induced by direct stimulation of the relevant areas of the brain (Penfield 1975). Penfield stimulated the motor cortex of conscious patients whose brain was exposed under conscious sedation. He found that the stimulation could produce complex, multistage, movements that appeared to be voluntary. The subjects, however, reported that they did not feel as if it was they doing the action. Daniel Wegner (2003) interprets this finding as showing that there is dissociation between conscious willing and voluntary action because the apparently voluntary action of Penfield's patients was not accompanied by the feeling of conscious willing. Wegner argues that this makes sense only if the experience of will is merely an addition to voluntary action.

There are a number of ways to challenge this kind of interpretation of Penfield's experiment. Wegner argues that the action appeared voluntary. But if our definition of voluntary behavior requires that an action is endogenous and is accompanied by conscious willing, there are two reasons not to classify the actions of Penfield's subjects as voluntary. The actions were not endogenous; they were externally caused by electrical stimulation. They were also not accompanied by the feeling of conscious willing, which is not surprising, given that they were not doing any willing. It seems, in fact, quite continuous with the common notion of free will that the patients induced to move by direct stimulation would not experience Libet's W. The reports of the subjects could be interpreted as evidence that the action was not voluntary. Hence, given that both of the elements of the common notion of free will are missing, it is not clear what evidence there is for Wegner to categorize those actions as voluntary. And it is this assumption that the action of Penfield's subjects were voluntary that provides the basis for the claim that the experiment constitutes evidence against the common notion of willing. It seems quite contrary to Wegner's claim that the subjects were able to identify a difference in subjective experience between the movements caused by direct stimulation of the brain and the everyday experience of willed motor movements. The fact that they could tell the difference could be taken as evidence that our subjective experience of willing is a veridical representation of the causal role of conscious experience in willing.

Further, purported evidence of the dissociation of conscious willing from the brain includes experiments utilizing transcranial magnetic stimulation (TMS) (Brasil-Neto et al. 1992). In this study, TMS was applied to either the left or the right motor cortex, again with the intent to influence movement of either the left or the right finger. Participants in this study were not able to identify the influence of TMS on their movements; instead they reported feeling as if they were willing to move either the right or the left finger. This study seems to lack the problem of Penfield's study in that the participants were convinced that they were the cause of their motor movements and not the TMS. However, unlike Libet's experiment, the TMS experiment does not study conscious willing as it might happen in close to normal circumstances.

3.3 Interpreting the Evidence against Free Will

Let us evaluate the claim that the studies presented in the previous section show that the concept of free will is erroneous. In order to evaluate this claim, one would have to identify exactly what is meant by free will. Libet proposes his own operational definition, which he thinks reflects the common notion of free will, i.e., the commonsense concept of free will. Libet's characterization of the common notion of free will includes the requirement of the conscious will. This way of characterizing free will is not unusual, and others have similarly established a connection between free will and consciousness. Patrick Haggard (2005) argues the Cartesian view that mental states cause the movements of the body is entrenched in our folk-psychological conception of voluntary action. George Sher (2009) argues that the view that free will and moral responsibility require consciousness is prevalent in philosophy as well.

Before I begin assessing the claim that the evidence presented in the previous section shows that the common notion of free will is false, I wish to carefully circumscribe the purpose of my argument. I do not plan to generate a definition of free will. Rather, I wish to evaluate the claim that evidence against conscious willing is evidence against the commonsense concept of free will. To evaluate that argument, I will rely on positions about common sense I have described in Chap. 2.

Furthermore, my argument in this section should be taken to apply only to the concept of free will and not to moral responsibility and culpability. It is true that free will is often thought to be the precondition for both, but what is required to impute moral responsibility or culpability is different from what is needed for attributions of free will. My argument is only about the requirement that consciousness is essential to the commonsense concept of free will. An argument about whether consciousness is needed for moral responsibility or even culpability is not required for my claims about free will.[3]

As I have argued in Chap. 2, commonsense concepts require the endorsement of a theory, which makes it more difficult to distinguish between commonsense and scientific theories in any way other than in terms of their quality. Furthermore, I argued that it might be difficult to circumscribe commonsense concepts because the quotidian notions we utilize are influenced by a variety of sources, including science. I argued that insofar as our commonsense concepts are influenced by science, their purported incompatibility with it is diminished. Finally, because our commonsense concepts are influenced by science it becomes difficult to identify exactly which, if any, concepts can be properly attributed the status of common sense.

The arguments about scope can also challenge the view that studies examining commonsense concepts are likely to yield definitive answers about common sense concepts as such. If it is true that historical context and cultural differences can influence how we interpret and predict human behavior, then any sampling of commonsense platitudes will have limited generalizability. If quotidian notions about concepts such as free will change with time and context, then studies about the common notions of free will can provide at best insight about current common sense. Claims about the scope and commitments of common sense are not often qualified. Rather, studies purport to capture enduring commonsense concepts. Hence, that our common notions are influenced by culture or that they change over time is a challenge to the claim that there is a definitive answer as to the commitments of common sense. Furthermore, a flaw in the commonsense concept of free will is not an argument against the existence of free will. Even if it were the case that the commonsense concept of free will is false, this is not enough to disqualify alternative accounts for the phenomenon of free will.

To dispute the claim that empirical evidence presented in the previous section shows that the common notion of free will is false, I will identify examples in everyday layperson ascription of free will that do not rely on consciousness. Again, my examples are not employed in an attempt to formulate an alternative definition of the common notion of free will, i.e., that free will does not require consciousness. It is an argument against the claim that Libet's construal of the common notion of free will captures all there is in our everyday parlance about free will.

There are examples of quotidian ascriptions of free will where there is dissociation between volitional action and conscious willing.[4] Many of them pertain to

---

[3] For a view that moral responsibility does not require conscious willing, see Smith (2005). For a critique of this view, see Levy (2013).

[4] The first few examples I have selected are of routine motor actions such as the ones selected by Libet. There is a dissanalogy to be made between volitional actions and volitional evaluative judgments required for moral deliberation. Smith (2005) argues that the way free will is defined by those who focus on its role in volitional action is not suitable for discussions of an individual's

overlearned automated behavior.[5] Imagine, for example, the action of tying one's shoelaces. Many people remember the first time they learned to tie their shoes, and recall that the learning process was a result of a number of deliberate, putatively conscious actions. The child needs to learn the separate elements involved in tying shoelaces and consciously attend to the performance of each. To aid in that process, there is even a song that is designed to help a child memorize the various stages of tying shoelaces. After a lot of practice, the child becomes expert at tying shoelaces and the process becomes automatic. When a practiced adult ties shoelaces, the process is entirely automated and does not require the adult to attend to any of the discrete movements necessary. Some of us can even attend to and complete additional tasks, such as reading the paper, while successfully tying shoes. It would be awkward, however, to argue that a person capable of reading the paper while tying her shoelaces is not performing a volitional action because she is not consciously willing each aspect of the process of putting on shoes.

The judgment that the person tying her shoelaces while reading the paper is still willingly performing that action is rooted in the presumption that automated processes, like tying one's shoes, often retain other elements of volitional action. As noted earlier, some of the often presumed elements of free will are that the action be endogenous and that the person be capable of selecting that particular action out of a number of alternatives. Both those elements of free will can be properly attributed to the person tying her shoes. She has decided to leave the house and put on a particular pair of shoes. Her decision was, relatively speaking, accomplished without external influence. She decided to put on that particular pair of shoes herself. She was not forced by physical means or verbal coercion. She presumably could have chosen to wear a different pair of shoes or not to tie her shoes. These two elements taken together lead to the conclusion that she willingly tied her shoes. Perhaps this also might mean that the other two elements of free will are prioritized over the element of conscious willing. At the very least, the fact that the attribution of free will is sometimes dissociated from the attribution of conscious willing is an indication that consciousness is not always a necessary element of the common notion of free will.

There are other examples of even more complex activities that are accomplished automatically, yet are likely to elicit an attribution of free will. Such activities

---

ability to choose a type of evaluative judgment required for moral deliberation. This is an argument against the suitability of Libet's experimental situations for showing that individuals lack free will in situations that involve moral deliberation. I agree with this criticism, but in my current discussion, I aim to challenge Libet's characterization of free will, and I am using examples of volitional action, rather than examples of volitional moral deliberation.

[5] Yaffe (2012), who discusses voluntariness in relation to legal culpability, argues that the law does not capture some instances of quotidian attributions of volition. He states that instances of bodily movements guided by unconscious mental representation would be commonly interpreted as voluntary in everyday parlance (p. 175). He also argues that habitual actions, even when complex, are in fact considered willed (and criminally liable), even when they are not intentional (p. 177). This gives further credence to the claim that there are cases of quotidian ascriptions of willing that are not conditioned on consciousness.

include many sporting activities, as well as expert playing of musical instruments. A particular conspicuous example is the game of tennis. Playing tennis illustrates how one can be involved in a complex motor task with very little reported awareness of decisions made about those motor movements. For anybody who has ever played or watched this fast-paced sport, it is clear that the ability to return the ball cannot be based on the conscious seeing of the ball and then the conscious willing to hit the ball in a particular way.[6] A study by Rob Gray (2004) demonstrated that batting performance by expert baseball players actually deteriorated if they were forced to focus on elements of skill execution, in this case direction of bat movements. Moreover, their performance did not suffer if they were asked to accomplish an additional task that would prevent them from focusing on batting. The opposite was true of less-expert players, which is expected if one assumes that conscious willing is required for a type of behavior to be learned. Novice baseball players performed better if they were able to pay attention to how they were directing the bat.

Despite this evidence that conscious willing is not required for expert sports performance, it would be peculiar at best to say that expert players are not acting freely when they are playing the game. Sports aficionados tend to stratify players in terms of the quality of their game and even evaluate the quality of the strategy involved in the game. This too is an example where quotidian ascriptions of free will are dissociated from conscious willing. As in the case of the woman who manages to tie her shoelaces without paying attention to her actions, the attribution of volition to the person comes from the presence of the other two elements of free will, which are that the action is endogenous and not coerced.

A challenge to my counterexamples could be that we attribute free will to the woman tying her shoes or to the expert baseball player because we presume that conscious volition was involved at the outset of the action.[7] It need not be the case that each physical action required for tying shoes must be caused by a conscious volition, but the initial decision to tie one's shoe needs to be conscious. It could be that judgments of free will are compatible with there being some unconscious mental activity that supports our actions just as long as it is the case that the decision that initiated the cascade of physical movements required to tie shoes is a conscious one. This preserves the primacy of mental causation and may perhaps account for instances of ascription of free will in cases where consciousness is not present for the duration of the volitional behavior.

To assess this argument, I will present evidence of the influence of unconscious processes on our behavior, which has been shown using the subliminal prime technique. A subliminal prime, or a masked prime, is an image, such as a word or a picture, presented below the threshold of consciousness. This technique has been used to demonstrate that subliminally presented primes are unconsciously perceived and that they influence behavior. For example, in one experiment, participants were asked to look for the appearance of a word on a computer screen (the target word), and then decide whether the word designated an artifact or a living thing. It turned

---

[6] For evidence for the claims about tennis, see Marcel (2003).

[7] I wish to thank an anonymous reviewer for this comment.

out that the categorization of the target word improved when preceded by a subliminal prime that was the same word. Thus, if the subliminal prime was *radio* and the
target word was also *radio*, the participants were able to more quickly categorize the
target as an artifact. When the prime was incongruous, for example, pairing *house*
and *radio*, the categorization was slower. This showed that although the participants
were not able to consciously perceive the prime, they were nonetheless influenced
by it when it came time to categorize the consciously perceived target word. When
the experiment was replicated using brain imaging, it was confirmed that subliminal
primes were not only perceived, but that they were processed deeply in the brain,
even in areas traditionally associated only with conscious processing (Dehaene
2014, p. 58).

Similar techniques were used to show that a number of other perceptual processes previously thought to require consciousness can operate subliminally, for
example, the binding of individual elements of a visual scene, such as pairing shape
and color or letters into words. Even multisensory information can be unconsciously
coupled. This is demonstrated by the "McGurk effect," where visual information of
an individual's mouthing the sound *ga*, coupled with the auditory stimulus of the
syllable *ba*, will produce the conscious perception of the syllable *da*. The person
hears *da* even though neither the auditory nor the visual stimulus corresponds with
that syllable. The explanation for the phenomenon is that through an unconscious
process the brain binds the two incongruous stimuli into a compromise between *ba*
and *ga*, which is *da* (Dehaene 2014, p. 62).

Even further, there is evidence that what becomes conscious is often prescreened
by unconscious attention. If attention is defined as a sifter that is required to distinguish relevant from irrelevant information when attending to a task, then there is
evidence that attention can operate unconsciously. For example, a subliminal prime
can attract attention to a particular location in a visual field, which will in turn
improve the ability to attend to consciously presented stimuli in that same location
(Dehaene 2014, p. 75). Furthermore, there is evidence that the conscious intention
can influence what we unconsciously attend to. When two stimuli are subliminally
presented, for example, a square and a circle, the intention to look for squares will
focus attention on the square even if the shape is not consciously perceived. That the
stimulus is processed is indicated by the heightened activation in the parietal lobe
(Dehaene 2014, p. 76).

Finally, there is evidence that unconsciously perceived signals can inhibit automatic responses. The ability to inhibit automatic responses is said to need the activation of the central executive system of the brain thought to require consciousness.
Participants asked to perform a repetitive task, for example, clicking a key whenever
a picture appeared on a screen, were able to inhibit that response when a stop signal
was presented, say a picture of a black disk. Surprisingly, the stop signal had an
inhibitory effect even when it was presented subliminally. Participants were able to
stop performing the repetitive task on cue even when it was only unconsciously
perceived (Dehaene 2014, p. 85).

The evidence for the role of unconscious processes in perception, binding, attention, and even their role in halting an automated process should lead us to conclude

that most volitional behavior will require the support of unconscious brain process-ing even in cases where aspects of the behavior are consciously willed. Consider the lady tying her shoes. Based on the argument for the primacy of conscious causation, her action is willed because at the outset she consciously willed to tie her shoes. The problem with this argument is that once we countenance that unconscious processes might be required in order to form a volition, the isolation of consciousness as the only cause of the action is unsupported. After all, one must be in the right context in order to form the volition to tie one's shoes: one must know that one is in a place where shoes are available; one must know the time of day and the location of the shoes; and so forth. The identification of the right context for tying shoes is no doubt accomplished in part through unconscious perceptual and attentional processes cap-tured by research utilizing subliminal primes. And if that is the case, then there are unconscious processes that play a causal role in the formation of the conscious voli-tion to tie shoes. Furthermore, if unconscious processes can cause conscious voli-tions, it is not obvious why we would pick out only the conscious state as the primary impetus for the action. All the states relevant to the formation of the con-scious volition can be properly said to be part of the causal chain that resulted in the action. The conscious volition was neither first in the causal chain nor uncaused by other states. Even further, based on the evidence presented, conscious processes are not required to inhibit an automated behavioral sequence. The unconscious pro-cesses involved in the formation of the conscious will are similarly necessary for the accomplishment of the task of putting on shoes. Thus, the reason to select the con-scious mental state as the only relevantly causative state is the commitment to the view that free will requires consciousness. Without that view, the conscious volition is just one among many factors that can influence behavior, and the primacy of the conscious volition cannot be established.

Perhaps one could argue that the view of the primacy of conscious volition can be reprised by arguing that although unconscious processes undergird conscious voli-tion and even make it possible to have volitions, the content of the volition, i.e., deci-sion to do or not to do something, has to be conscious, much like Libet's veto power of the conscious will. The research by Soon et al. (2008), however, supports the claim that conscious volition is sometimes preceded by unconscious activity that deter-mines the conscious volition to move the hand in a certain way. Thus, it is not just that unconscious brain processes supply the groundwork necessary for a conscious volition; it is that the unconscious process determines the content of the decision. This result is a challenge only if one is committed to the view that in order for some-thing to be volitional, it has to be conscious. If one countenances the possibility of unconscious volitions,[8] the result established by the Soon at el. (2008) study shows at most that the common notion of free will as formulated by Libet is inadequate.

Initially, it might seem a category mistake to call anything an unconscious voli-tion, but research using subliminally presented information provides evidence that such volitions do occur and influence behavior. For example, in a study measuring the influence of subliminal incentives, participants were asked to squeeze a handle after seeing an image of a monetary reward they were likely to receive if they

---

[8] For more on unconscious volitions, see Rosenthal (2002).

squeezed hard enough. The monetary rewards varied in amount. It was shown that participants squeezed the handle harder for the higher reward, even in trials where the image of the money was presented subliminally (Dehaene 2014, p. 77). In other words, participants worked harder to obtain the higher reward even when they were not consciously aware of what was at stake. In order to react to the unconscious stimulus, the participants must have formed something akin to the volition to squeeze harder for the higher reward, and yet neither the incentive nor the decision to squeeze harder was consciously willed. Moreover, that they squeezed harder for the higher reward is congruous with what would likely be most individuals' conscious response.

Circling back to the issue of the compatibility between commonsense concepts and scientific challenges to those, there is evidence that neuroscientific explanations of human behavior are perceived as undermining our folk beliefs about free will (Nahmias et al. 2007). Further research, however, clarifies that what is most important for folk attribution of free will and moral responsibility is that the action be caused by a mental state. According to Murray and Nahmias (2014), the folk think that neuroscientific explanation is incompatible with free will only in scenarios where there is *bypassing*, i.e., neuroscience is construed as supplanting or omitting mental causation. In cases where mental causation is accounted for by neuroscience, attributions of free will persist, and neuroscientific explanations and free will are seen as compatible (Murray and Nahmias 2014). None of the recounted evidence for the presence of unconscious mental states is evidence against mental causation. On the contrary, it is an argument for the inclusion of unconscious states into the realm of the mental.

3.4 Common Sense and Determinism

Libet (1999) distinguishes two distinct problems for free will. The first is the phenomenon demonstrated by his studies, and subsequently confirmed by others, that unconscious brain activity seems to play a role in volitional behavior. Libet, however, says that there is a distinct problem of reconciling free will with determinism, which his studies do not directly address. I will now focus on the issue of scientific determinism and free will, with an emphasis on the purported conflict between the commonsense concept of free will and scientific determinism.

The difficulty in reconciling determinism and free will can be explained in the following way. As neuroscience continues to advance, we will be able to explain more of human psychology in terms of brain activity. This expansion of the scientific domain has been interpreted as particularly troubling for free will because it augments the threat of scientific determinism. Determinism is the claim that, given a certain set of initial conditions (for example, conditions that existed at the time of the Big Bang) and given the laws of physics, which specify a cause for each event, every event from the onset of the universe can be explained and predicted. Now, if psychological processes can, in some as yet unknown way, be subsumed under the laws of physics, the laws of physics will determine human psychology and each human action. It would be false, then, to say that persons are free to make choices; in the same way, it would be false to say that a ball falling from a height has the choice to obey the law of gravity. A decision one makes is caused by events preceding that decision, and those events in turn were caused by events before them, and so on, forming a long causal chain that reaches all the way back to the beginning of the universe. Based on this picture, things could not have been other than they are and any individual person could not have done otherwise.[9]

I will now assess attempts to utilize commonsense notions in the debate about free will and scientific determinism. There are those who argue for the elimination and yet others for the inclusion of common sense into the development of the concept of free will. I will divide commentators into two groups, the friends and the foes of common sense. I will argue that both camps assume that there is a settled commonsense view. The foes argue that this settled commonsense view is erroneous and should be eliminated, while the friends argue that common sense should be the basis of a more developed view of concepts such as free will, determinism, and moral responsibility. Both camps, I argue, mischaracterize common sense.

Let us begin with the foes of common sense. Greene and Cohen (2004) describe the commonsense concept of free will as libertarian and therefore incompatible with scientific determinism. Green and Cohen define libertarianism as the view that scientific determinism is false (Green and Cohen 2004, p. 1776). A study by Nichols and Knobe (2007) supports the argument that we are libertarians when it comes to people, but determinists when it comes to physical processes. They found that when asked whether physical events, say, water coming to a boil, are entirely determined, study participants responded affirmatively. However, participants disagreed that volitional action, such as stealing a candy bar, was predetermined.

Greene and Cohen argue that it is much easier to maintain a libertarian point of view in the face of abstract and general arguments about determinism, perhaps like the ones used in the Nichols and Knobe study. It is, however, much harder to maintain such views as science is turning the black box of the mind into a "transparent bottleneck." Greene and Cohen argue that the brain is that bottleneck through which all of the causes from the past contribute to who we are today, and neuroscience is making the workings of this bottleneck transparent: "At some time in the future we may have extremely high resolution scanners that can simultaneously track the neural activity and connectivity of every neuron in a human brain. ...Imagine, for example, watching a film of your brain choosing between soup and salad" (Greene and Cohen 2004, p. 1781). Implicit in this view is the assumption that as brain processes are revealed and their connection with human behavior clarified, there will be less reason to believe in the type of mental causation required for attributions of free will.

Greene and Cohen maintain that, given the success of neuroscience in making brain function transparent, we will have to eliminate our commonsense concept of

---

[9] The assessment of the threat of determinism for free will is beyond the scope of this chapter. For more on this issue, see Gligorov (2012).

free will. They argue both that the commonsense concept of free will is libertarian and that progress in neuroscience will make it obvious that determinism is true. They notice, however, the slack between what we might believe to be true and how we perceive the world and other people and argue that perhaps free will is an incorrigible illusion. Consider perceptual illusions: We all know that straight objects put under water look bent because of refraction. Nonetheless every time you place a pencil into a container full of water, the pencil will continue to appear bent despite your knowledge that the pencil is not bent. Perhaps attributions of free will are just like that—we cannot help but attribute free will to people: "Having learned from experience and by reasoning that a stone falls downward, man is convinced beyond doubt and in all cases expects to find this law operating which he has discovered. But having learned just as surely that his will is subject to laws, he does not and cannot believe it" (Tolstoy ed. 1982, p. 1473).

In case common sense about free will turns out to be an illusion, then complete replacement of the commonsense concept of free will is required to make space for the scientific conception of how humans direct their actions. Characterizing free will as an illusion, however, is contradictory to Greene and Cohen's description of the impact of neuroscience on the maintenance of commonsense beliefs. It is not clear why the progress of neuroscience, which is, as they say, turning the brain transparent, should make any difference to the conception of free will if it is an illusion.

In principle, one could support the push to eliminate the commonsense concept of free will using Sellars's argument, presented in Chap. 2, that folk psychology is a theory that predicts and explains human behavior by positing internal states (Sellars 1997 ed.). Based on Sellars's view, change in theory will result in a changed conceptual framework, which in turn would result in a different way of reporting our inner states. Hence, if a better theory about human psychology were to replace FP, and if that theory did not feature free will, our everyday predictions and explanations of human behavior would lack reference to free will as well. But if free will is akin to a perceptual illusion, then no change in theory would correct this illusion because perceptual illusions result from biological facts about human perception. However, attributions of free will are sufficiently different from perceptual beliefs, and the analogy between those two should be questioned.

In addition to Greene and Cohen, Wegner (2003) characterizes free will as an illusion, but he further claims that the concept of free will is an inference to the best explanation of human behavior and that it is an empirical matter as to what degree humans are free (Wegner 2003 68). If that is the case, then certainly if there is a revision to the best explanation, it would result in the change of the concept of free will. Perhaps it might turn out to be the case that there are facts about human psychology that make it impossible to interpret human behavior without attributing free will. But this seems unlikely, given that, as I will show, humans are capable of adjusting their attribution of free will based on facts about human psychology.

The need for replacement of (FP) rests on the purported incompatibility of the commonsense notion of free will, characterized here as libertarian, and scientific determinism. The ontology of FP can be made to be incompatible with that of neuroscience if one accepts that FP comprises only nonscientific platitudes, an argument I disputed in Chap. 2. Scientific facts about human psychology are already incorporated into our everyday parlance, as in the general awareness that the person who has a neurological disorder, say, Tourette's syndrome, has diminished control over his or her actions. Even our understanding of psychiatric diseases makes us more hesitant to blame those who have psychiatric conditions. For example education about the biological etiology of some mental illness can affect interpretation of the degree of control individuals can exert over their behavior (Boysen and Vogel 2008). Although psychiatric illness is still stigmatizing, the stigma is diminishing because of the expansion of the scientific understanding of human psychology and human behavior. The classification of some conditions as medically treatable conditions encourages the reinterpretation of those as disorders with physical causes. For example, the recasting of depression as a condition with physical causes most likely comes from the finding that treatments for it are efficacious by influencing the chemistry of the brain. Goldstein and Rosseli (2003) showed that a biological model of depression was associated with a decrease in types of stigma. If a condition can be either identified with a neurological cause or treated by changing aspects of brain functioning, we are more likely to accept it as a medical condition and accept that the individual has diminished free will. Based on this more congenial construal of FP, the quotidian notion of free will adjusts to relevant scientific advances.

Now I will turn to consider the friends of common sense, who dispute the view that determinism is incompatible with FP. Nahmias (2006) argues that determinism is not relevant to judgments about moral responsibility. Instead, Nahmias contends, free will is undermined by the reduction of mental to physical states. In his study, participants were asked to assign praise or blame after reading several different scenarios. Nahmias (2006) found that people were less likely to hold somebody responsible in scenarios where they were told that reduction of mental to brain states was achieved.

In a further study by Murray and Nahmias (2014), also discussed in Sect. 3.3, participants judged determinism compatible with free will, if it was explained that determinism would not entail bypassing or epiphenomenalism (the view that mental states do not have causal powers). In scenarios that described individuals living in a determinist world, where there was still mental causation, participants continued to ascribe free will. Another way of looking at these results is that Murray and Nahmias corrected for the misconception that determinism precludes mental causation, and that reconceptualization of determinism yielded a different conclusion about the compatibility of free will and determinism.

Murray and Nahmias explain that they applied Frank Jackson's (1998) method of possible cases to elicit more accurately intuitions about the concepts being investigated. Participants responding to cases having in mind an erroneous definition of determinism—one that does entail bypassing mental causation—might report false intuitions, not indicative of the commonsense concept under investigation. According to Jackson, our first judgments are not usually the most accurate ones (Jackson 1998, p. 35).

The reason to strive to determine the extent of folk intuitions, according to Murray and Nahmias, is to help achieve reflective equilibrium about the concepts of free will and determinism. Reflective equilibrium, they maintain, is the default method for theorizing about normative concepts, such as free will: "…wide reflective equilibrium (WRE), takes as inputs our normative principles, background scientific theories, and pre-theoretical (but reflective) judgments, or intuitions about relevant cases, and then attempts to develop a philosophical theory that is maximally consistent (and, ideally, mutually justifying) among those inputs (Murray and Nahmias 2014, p. 435)."

There are two elements of this argument I wish to challenge. The first is the claim that there are such things as pre-theoretical judgements. The second element is the argument that the process proposed by Jackson and adopted by Nahmias is likely to yield commonsense or folk beliefs about any particular phenomenon. I will argue that the considered judgments captured by the studies are not common sense.

In Chap. 2, I described Lewis's method for defining the limits of commonsense psychology. He argues that we should collect all the relevant platitudes we use to ascribe mental states and those we use to predict people's behavior. The collection of those platitudes will contain both theoretical and observational terms. Observational terms are those that are defined by other theories, while theoretical terms are those implicitly defined by folk psychology. The collected platitudes specify the functional role of the theoretical terms. The theoretical terms of folk psychology are mental terms, and all the others are observational terms. Lewis (1972) argues that the theoretical terms of folk psychology are pre-theoretical because they are introduced by commonsense psychology.

The distinction, as Lewis draws it, does not support the argument that there are beliefs or platitudes that employ terms that could have meanings without endorsing a theory. Mental terms are pre-theoretical because they are introduced by commonsense psychology, but the meanings of those terms are set by the role they play in that theory. As Lewis further argues, theoretical terms can in principle be eliminated in favor of the role specified by observational terms. Thus, it makes no sense to argue that there are meaningful terms that are not part of at least a minimal background theory. The scope of such a theory could be specified by collecting the relevant platitudes that feature the term in question, say, free will or determinism.

Sellars's argument, as presented in Chap. 2, Sect. 2.2 supports a similar conclusion. Sellars's argument is that utilization of any term, including the term free will, requires familiarity with a background conceptual framework. Thus, although our platitudes might express judgments about the applicability of certain terms, those judgments are not pre-theoretical in any meaningful way. This is important because pre-theoretical terms are often treated as an evidentiary basis upon which a theory of a particular concept could be built, when in fact they should be treated as expressing the endorsement of a particular view already.

I will now argue for why I think the results of the Murray and Nahmias study can be interpreted to support the argument that what happens to folk intuitions is not, as they claim, refinement of common sense, but a change in background conceptual commitments. The participants in their study were asked to decide whether Bill, as described in the experimental scenario, has free will. In the scenario, Bill decides to steal a necklace in order to impress a coworker to whom he is attracted. Bill's universe is described as follows: "In Universe A every decision is completely caused by what happened before the decision. This does not mean that in Universe A people's mental states (their beliefs, desires, and decisions) have no effect on what they end up doing, and it does not mean that people are not part of the causal chains that lead to their actions" (Murray and Nahmias 2014, p. 451). The study participants overwhelmingly ascribed free will to Bill, which means that they judged free will and determinism to be compatible.

Murray and Nahmias (2014) argue that in previous studies the incompatibilist judgments by the study participants were elicited because the folk interpreted determinism to entail either epiphenomenalism or fatalism (p. 440). Those assumptions, then, led them to conclude that one cannot attribute free will to individuals in a deterministic universe. Teaching them an alternative definition of determinism resulted in changed beliefs. Based on Jackson's view, one could argue that the method of possible cases was used to crystalize commonsense judgments about determinism and its compatibility with free will, and the study by Murray and Nahmias (2014) showed that commonsense is compatibilist.

Alternatively, one could argue that the participants changed their mind because they were apprised of a different definition of determinism. The introduction of a different construal of determinism as well as the insertion of mental causation into a determinist universe altered the implicit definition of free will. In effect, the new view replaced fatalism or epiphenomenalism, which was the view that generated the purported folk intuitions in previous studies. Murray and Nahmias argued that introducing the new definition constituted a refinement of common sense. But if epiphenomenalism or fatalism was at the root of previous judgments about compatibilism, it is wrong to argue that replacing those views would constitute refinement of common sense. Thus, what is captured by the experimental scenarios presented by Murray and Nahmias is that there are definitions of determinism which are compatible with ascriptions of free will, not that commonsense is compatibilist.

My argument should not be misunderstood to support the view that earlier studies were more successful in capturing commonsense views about free will or that I think that fatalism or epiphenomenalism are common sense. As I have argued previously, once it is countenanced that the use of any concept requires at least the implicit endorsement of a conceptual framework, it becomes difficult to distinguish common sense beliefs from other types of beliefs. The argument I have put forth is that the introduction of mental causation into the definition of determinism is not a refinement of commonsense, but a demonstration that beliefs about the compatibility of free will with determinism are affected by how each of the relevant concepts is characterized.

3.5 Conclusion

In this chapter, I have presented several studies that have been interpreted as evidence against free will, particularly, Libet's experiments showing that conscious volition is preceded by unconscious brain activity. I have assessed the arguments that Libet's experiments are a threat to the common notion of free will. I dispute the idea that the common notion of free will requires conscious willing and I provide examples of automated behavior, where there is the tendency to attribute free will in the absence of consciousness. Furthermore, I substantiate the claim that unconscious processes play a role in the formation of conscious volitions. Based on that, I argue against the primacy of conscious mental causation. I conclude that none of the studies cited provide evidence against the phenomenon of free will, only an argument against Libet's characterization of the common notion of free will.

I also assess calls for the elimination of the commonsense concept of free will and conclude that this view rests on a particularly unfavorable characterization of free will. I argue that the concept of free will also cannot be properly characterized as an illusion. Moreover, I argue that attributions of free will are responsive to facts about human psychology in a way that indicates that a notion of free will can be adjusted to accommodate for scientific advances in neuroscience. Finally, I argue that any concept of free will requires endorsement of a particular background theory. Because of that I question whether any concept of free will can be properly characterized as common sense. To support the claim that background theory matters for judgments about the compatibility of the commonsense concept of free will and determinism, I propose a reinterpretation of the current evidence about folk attitudes toward the compatibility of those two.

References

Banks, W. P., & Ischam, E. A. (2009). We infer rather than perceive the moment we decided to act. *Psychological Science, 20*, 17–21.

Boysen, G. A., & Vogel, D. L. (2008). Education and mental health stigma: The effects of attribution, biased assimilation, and attitude polarization. *Journal of Social and Clinical Psychology, 27*(5), 447–470.

Brasil-Neto, J. P., Pascual-Leone, A., Valls-Solé, J., Cohen, L. G., & Hallett, M. (1992). Focal transcranial magnetic stimulation and response bias in a forced-choice task. *Journal of Neurology, Neurosurgery, and Psychiatry, 55*, 964–966.

Dehaene, S. (2014). *Consciousness and the brain: Deciphering how the brain codes our thoughts.* New York: Viking Penguin.

Gligorov, N. (2012). Determinism and advances in neuroscience. *American Medical Association Journal of Ethics, 14*, 489–493.

Goldstein, B., & Rosseli, F. (2003). Etiological paradigms of depression: The relationship between perceived causes, empowerment, treatment preferences, and stigma. *Journal of Mental Health, 12*(6), 551–563.

Gray, R. (2004). Attending to the execution of complex sensorimotor skills: Expertise differences, choking, and slumps. *Journal of Experimental Psychology: Applied, 10*(1), 42–54.

Greene, J., & Cohen, J. (2004). For the law, neuroscience changes nothing and everything. *Philosophical Transactions of the Royal Society, B: Biological Sciences, 359*(1451), 1775–1785.

Haggard, P. (2005). Conscious intention and motor cognition. *Trends in Cognitive Science, 9*(6), 290–295.

Jackson, F. (1998). *From metaphysics to ethics: A defence of conceptual analysis.* New York: Oxford University Press.

Lau, H. C., Rogers, R. D., & Passignham, R. E. (2007). Manipulating the experience onset of Intention after action execution. *Journal of Cognitive Neuroscience, 19,* 81–90.

Levy, N. (2013). The importance of awareness. *Australasian Journal of Philosophy, 91*(2), 211–229.

Lewis, D. (1972). Psychological and theoretical identifications. *Australasian Journal of Philosophy, 50*(3), 207–215.

Libet, B. (1999). Do we have free will? *Journal of Consciousness Studies, 6*(8–9), 47–57.

Libet, B., Wright, E. W., & Gleason, C. A. (1982). Readiness potentials preceding unrestricted spontaneous pre-planned voluntary acts. *Electroencephalography & Clinical Neurophysiology, 54,* 322–325.

Libet, B., Gleason, C. A., Wright, E. W., & Pearl, D. K. (1983). Time of conscious intention to act in relation to onset of cerebral activity (readiness potential): The unconscious initiation of a freely voluntary act. *Brain, 106,* 623–642.

Marcel, A. (2003). The sense of agency: Awareness and ownership of action. In J. Roessler & N. Eilan (Eds.), *Agency and self-awareness* (pp. 48–93). New York: Oxford University Press.

Murray, D., & Nahmias, E. (2014). Explaining away incompatibilits intuitions. *Philosophy and Phenomenological Research, LXXXVIII*(2), 434–467.

Nahmias, E. (2006). Folk fears about freedom and responsibility: Determinism vs. reductionism. *Journal of Cognition and Culture, 6*(1–2), 215–237.

Nahmias, E., Coates, D. J., & Kvarian, T. (2007). Free will, moral responsibility, and mechanisms: Experiments on folk intuitions. *Midwest Studies in Philosophy, XXXI,* 214–242.

Nichols, S., & Knobe, J. (2007). Moral responsibility and determinism: The cognitive science of folk intuitions. *Noûs, 41*(4), 663–685.

Penfield, W. (1975). *The mystery of mind.* Princeton: Princeton University Press.

Rosenthal, D. (2002). The timing of conscious states. *Consciousness and Cognition, 11,* 215–220.

Sellars, W. (1977, 1997 ed.). *Empiricism and the philosophy of mind.* Cambridge, MA: Harvard University Press.

Sher, G. (2009). *Who knew? Responsibility without awareness.* Oxford: Oxford University Press.

Smith, A. M. (2005). Responsibility for attitudes: Activity and passivity in mental life. *Ethics, 115,* 236–271.

Soon, C. S., Brass, M., Heinze, H.-J., & Haynes, J.-D. (2008). Unconscious determinants of free decisions in the human brain. *Nature Neuroscience, 11*(5), 543–545.

Tolstoy, L. (1869, 1982 ed). *War and peace* (R. Edmonds, Trans.). London: Penguin Classics.

Wegner, D. W. (2003). The mind's best trick: How we experience conscious will. *Trends in Cognitive Science, 7*(2), 65–69.

Yaffe, G. (2012). The voluntary act requirement. In A. Marmor (Ed.), *The Routledge companion to philosophy of Law* (pp. 174–190). New York: Routledge.

Chapter 4
Cognitive Enhancement and Personal Identity

Abstract Enhancement can be defined as the improvement of normal individuals. There are several categories of enhancement, including physical enhancement, cognitive enhancement, and moral enhancement. In this chapter, I focus on the argument that cognitive enhancement using pharmaceutical means could cause disruptive changes in personal identity. I distinguish between numerical and narrative identity. I argue that cognitive enhancement would have no effect on numerical identity, but it could affect narrative identity. Narrative identity approximates the common notions of identity because it is characterized as a first-person effort to construct a concept of self. Despite the potential effect on narrative identity, I argue for the permissibility of the use of cognitive enhancers. I maintain that psychological traits can change without disrupting psychological continuity. This view is supported by evidence that individuals experience a great deal of psychological change over time, and the evidence that even when those changes are caused by the use of medication, they do not always create a disruption in narrative identity. I conclude that cognitive enhancement is permissible even when it produces changes in narrative identity.

4.1 Introduction

Enhancement can be defined as the improvement of normal individuals. There are several categories of enhancement, including physical enhancement, cognitive enhancement, and moral enhancement. Physical enhancement includes attempts to improve physical performance in sports as well as physical appearance through surgical means. Cognitive enhancement, sometimes also referred to as neuroenhancement, encompasses pharmaceutical methods of improving memory, concentration, and promotion of wakefulness. Finally, there is the category of moral enhancement. Moral enhancement relies on the possibility of the use of pharmaceutical agents, and any other medical technology, for the moral improvement of human beings. In this chapter, I primarily discuss cognitive enhancement.

Although the prospect of improved intellectual prowess seems appealing, there are moral issues to consider before adopting a permissive stance with regard to the use of cognitive enhancers. There are distinct ethical questions raised about the use of cognitive enhancers, which rely on the distinction between the use of medicine

© Springer Science+Business Media B.V. Dordrecht 2016
N. Gligorov, *Neuroethics and the Scientific Revision of Common Sense*,
Studies in Brain and Mind 11, DOI 10.1007/978-94-024-0965-9_4

for treatment and its use for enhancement. Sandel (2004) argues that the only acceptable application of medicine is for treatment of disease and prevention of disability. All other uses of medicine and medical technology are outside of the scope of this traditional conception and are not morally permissible. Assuming that there is a distinction between treatment and enhancement, the balance between risks and benefits of using neuroenhancers could be construed as unfavorable (Chatterjee 2004). If medication is used for the treatment of disease, the risks of having a disease outweigh the risks of potential side effects of treatment. This balance is usurped if drugs are used solely for improvement of normal individuals, not exposed to the deleterious effects of disease. The distinction between treatment and enhancement, however, is not accepted by everyone. For example, Norman Daniels (2000) argues that those two applications of medicine are not easy to distinguish and that the line between them often becomes blurred. I agree with Daniels and have argued against the distinction between treatment and enhancement (Gligorov 2010).

Additional concerns about cognitive enhancement are raised by M. J. Farah et al. (2004); they discuss the possibility of forced use of neurocognitive enhancers. Employers might recognize the benefits of enhancers on productivity, including improved memory, the prolonged ability to concentrate on particular tasks, and the increased ability to stay awake. They might disregard any risks of the use of neuroenhancers and require employees to use them regularly. Farah et al. (2004), and Glannon (2007), note the potential problem for distributive justice as well. People who can afford medicine will easily avail themselves of performance-enhancing drugs while those of more modest means will not be able to partake in their benefits. The potential widespread use of neuroenhancers, especially to improve academic performance, might raise the average performance of students in such a way that those who do not have access to cognitive enhancers will be permanently disadvantaged and unable to compete.

Finally, some have warned of the possibility that neuroenhancers could affect personal identity and authenticity. These concepts have been implicated in the debate about cognitive enhancers because change in cognitive abilities might result in a changed self. I will focus this chapter mainly on a discussion of personal identity. In Chap. 5, I will assess the impact of memory modification on authenticity.

This chapter is divided into two parts. In Sect. 4.2, I list some pharmaceutical agents that have been identified in the literature for their potential use as cognitive enhancers. In the section, I aim to provide the reader with the background against which to judge claims about the perils of cognitive enhancement as well as to illustrate that the possibility of neuroenhancement is still in the future.

In Sect. 4.3, I discuss the purported effects of cognitive enhancement on personal identity. I argue that there are two distinct concepts to consider, numerical identity and narrative identity. I argue that numerical identity will remain unaffected by the use of cognitive enhancers, while narrative identity, which approximates the common notions of identity, might be affected. I argue further that use of enhancers might precipitate a change in psychological traits, but such changes will not always lead to a disruption in narrative identity. Thus, I argue that use of cognitive enhancers is permissible even if it leads to changes in psychological traits. However, my argument leaves open the possibility that individual instances of neuroenhancement could be impermissible, for example, in cases where the negative side effects to an individual outweigh the benefits.

4.2   Potential Cognitive Enhancers

Pharmaceutical agents most often characterized as cognitive enhancers include stimulants, such as methylphenidate (Ritalin®) and dextroamphetamine (Adderall®). The Food and Drug Administration (FDA) has approved both of those drugs for the treatment of individuals with Attention Deficit and Hyperactivity Disorder (ADHD). For normal individuals such stimulants have been shown to increase concentration and improve performance on cognitive tasks as well.

The purported effectiveness of stimulants in improving cognitive performance has had an impact on the prevalence of their use. A number of surveys show that Ritalin® and Adderall® are increasingly being used by college students, either legally or illicitly, for the purpose of improving academic performance (Advokat et al. 2008; Hall et al. 2005). Moreover, stimulants are prescribed off-label, for an FDA-unapproved indication. Using a survey, Advokat et al. (2008) concluded that about 57 % of students who reported using stimulants did not have a diagnosis of ADHD, but obtained stimulants legally by having a prescription. Despite this uptick in the use of stimulant drugs, their efficacy for the purpose of improving academic performance remains questionable.

There are a few studies of methylphenidate and dextroamphetamine, testing their effects on normal individuals, and those show limited increase in aspects of cognitive performance. Elliot et al. (1997) studied the effects of methylphenidate in 28 healthy male volunteers. The trial was a double-blind, placebo-controlled study, with the volunteers randomly assigned to receive either a placebo or methylphenidate. The participants were given a battery of tests, including tests for spatial working memory, planning, verbal fluency, and attention. Methylphenidate was shown to improve spatial working memory on some tasks, although the improvements were only seen when the task was novel. In repeated performances of this task, the stimulant seemed to be detrimental to performance, and individuals who did not take the drug performed better on the task of spatial working memory. The drug had no effect on verbal fluency or attention (Elliot et al. 1997).

An additional study by Mehta et al. (2000) showed methylphenidate to produce improvements in working memory. Those were most prominent for individuals who started with a lower baseline of working memory prior to the administration of the drug. This study seems to contest the worry expressed in Farah et al. (2004) that use of cognitive enhancers would improve cognition for everybody, thereby creating further inequality in education and access to opportunities in general. Stimulants could increase the average performance on certain cognitive tasks, but by improving the performance of those identified as low performers.

In a study by Izquierdo et al. (2008), methylphenidate was shown to improve long-term memory in individuals who were over the age of 35. The drug was not shown to improve encoding, the amount of information learned, but it improved recall, 7 days after the learning experience. The drug had no beneficial effects on long-term memory for individuals younger than 35 (Izquierdo et al. 2008). This study could undermine the status of Ritalin® as a study drug because its effects on memory seem to be prominent only in individuals who are older than the average college student. Moreover, the study indicates positive effects on recall a week after the use of the drug, which does not show anything about the efficacy of taking this drug immediately before an exam, as might be the practice of college students who are not regularly treated with stimulants.

Because of the scarcity of evidence about the efficacy of stimulants for enhancement of cognition in normal individuals, it is useful to include a discussion about the efficacy of these agents to improve cognitive performance in those with ADHD. ADHD is most often diagnosed in childhood. Children with ADHD have difficulty staying focused, paying attention, difficulty controlling behavior, and are hyperactive. In a meta-analysis of the overall efficacy of stimulants for treatment of ADHD, Advokat (2010) concludes that stimulants do improve classroom manageability and increase attention and productivity in children with ADHD. However, use of stimulants does not improve long-term academic accomplishment. Children with ADHD, despite treatment, have consistently lower IQ than normal controls. They still score lower on reading and arithmetic tests and require more remedial academic services. Children with ADHD: "take more years to complete high school, and have lower rates of college attendance and graduation." (Advokat 2010, p. 1257). Because of the lack of evidence for long-term academic improvement, Advokat (2010) argues that the classification of stimulants as cognitive enhancers is not warranted.

A further class of drugs with the potential for use as cognitive enhancers is acetylcholinesterase inhibitors, including donepezil, galantamine, and rivastigmine. Out of these three the one most studied for its enhancing properties in normal individuals is donepezil, which is FDA approved for the treatment of Alzheimer's disease and marketed in the U.S. as Aricept®. Yesavage et al. (2002) performed a randomized, double-blind study of 18 licensed pilots ranging from 30 to 70 years of age (with a mean age of 52) to determine the effects of donepezil on the retention of skills required for aviation. To do that they studied how the drug affected pilots' performance in a flight simulator. The pilots were trained for 75 hours in a flight simulator to perform a series of complex tasks. They were then given either donepezil or a placebo for the duration of 30 days. After that period, the pilots were retested, using the same set of tasks in the flight simulator. The pilots who were treated with donepezil retained their aviation skills at the same level as during the initial session, while those in the placebo group experienced deterioration in performance as measured against their baseline 30 days prior. The study showed that donepezil was efficacious in maintaining the ability to perform the set of complex flight-related tasks (Yesavage et al. 2002).

A meta-analysis by Repantis et al. (2010) reviewed a few other studies of donepezil using healthy adult subjects. The success of donepezil in improving cognitive performance was mixed. In one study cited by Repantis et al. (2010), donepezil was shown to improve episodic memory, but those results were not confirmed in another study of healthy individuals, who under conditions of normal wakefulness had no cognitive improvements. In a study by Beglinger et al. (2004), donepezil actually showed a negative effect on attention and verbal memory tasks in healthy young and older adults.

Another potential neuroenhancer is modafinil (Provigil®). It has been FDA approved for the treatment of narcolepsy, but has been prescribed off label for a variety of sleep disorders, including sleep apnea. Two studies on normal healthy adults showed that modafinil could be successfully used to abate the negative effects of sleep deprivation. In a study by Grady et al. (2010), healthy patients underwent a protocol in which the period of sleep-wakefulness was significantly different from their usual. The participants remained awake for longer and slept fewer hours. The study was a randomized double-blind, placebo-controlled study. The participants who received modafinil for the duration of the experiment were better able to remain awake and alert. Modafinil was particularly efficacious in improving cognitive-psychomotor speed and attention. Furthermore, a study of sleep-deprived physicians showed that modafinil was successful in diminishing the cognitive deterioration associated with sleep deprivation (Sugden et al. 2012).

4.3   Different Concepts of Identity

Several commentators, including Carl Elliot (1999) and the members of the Presidential Council on Bioethics (2003) have brought up alterations to personal identity as a potential moral obstacle to neurocognitive enhancement.[1] The charge is that the use of enhancers could change not just our intellectual abilities but might also alter core personality characteristics. The kinds of improvements mentioned in Sect. 4.2 do not seem to be immediately relevant to personal identity. One could envision, however, that improved ability to perform on certain cognitive tasks and perhaps become better at one's job could affect a person's self-image in a number of ways. For example, a person could become more confident, more ambitious, and even more social. Better performance in school or at work might lead to better jobs and higher socioeconomic status, all of which could improve self-image.

DeGrazia (2005a) aptly notes that in order to evaluate the claim that neuroenhancers can change personal identity, we need to make clear what is meant by personal identity. To do that, he distinguishes between two senses of identity, numerical identity and narrative identity. I will describe numerical identity first. Numerical identity may refer to either synchronic or diachronic identity, where synchronic identity is the identity of an object or individual at a time and diachronic identity is identity of an object or individual across time. In this section, the focus will be on identity across time, or diachronic identity. My exposition, however, is

---

[1] This problem is also mentioned in Glannon (2007) and in Farah et al. (2004).

not meant to be a comprehensive review of the vast literature on the topic of personal identity. I aim to provide only enough information on the topic to support my claim that when it comes to cognitive enhancers, we should focus on narrative identity.

One of the fundamental principles of identity is described by Leibniz's law, which states that two things are identical if they are qualitatively indiscernible, i.e., two things, a and b, are identical if any property of a is a property of b as well. For example, the car I drove to work yesterday is the same car I drove to work today if any property of yesterday's car is also a property of the car I drove to work today.

Additional features of identity include reflexivity, symmetry, and transitivity. Reflexivity presumes that each thing is identical to itself; the car I drove to work today stands in the relationship of identity to itself. Identity is also symmetrical; the car I drove to work yesterday is identical to the car I drove to work today and the car I drove to work today is identical to the car I drove to work yesterday. Transitivity of identity entails that if a and b are identical and b is identical to c, then a and c are identical as well. If the car I drove to work yesterday is the same car I drove to work today, and if the car I drove to work today is identical to the car I will drive to work tomorrow, then the car I drove to work yesterday is identical to the car I will drive to work tomorrow.

Given the above formulation of identity, it is easier to see why change over time might pose a problem for the maintenance of identity. Imagine that over time my car begins to need some repairs and parts of the car are replaced as they stop functioning. As each part of the car is replaced by a new one, the question of identity can be raised. If a new brake system is put in place, one could ask whether the car I drove before the change in brakes is the same as the one after. When the carburetor is replaced, one might puzzle over whether diachronic identity of the car is maintained. If over time every single part of the original car has been replaced, one might question whether the old car has ceased to exist and a new car has taken its place.[2]

The problem of identity of persons over time can be formulated in the following way: If a person's life is conceived of as truncated into distinct stages, for example, a stage at age 6, a stage at age 15, and a stage at age 35, identity would require that those three distinct stages of a person (or person stages) maintain the relationship of identity among each other.[3] Hence, the person stage at 6 and the stage at 15 would have to be identical, and identity would have to be maintained among those earlier stages of the person and the person's current stage at 35 years of age. Identity, if one chooses to use Leibniz's sense of identity, would also require that the relationship between the person stages obey symmetry (the person stage at age 6 is identical to the person stage at age 15 and the other way around), and transitivity (the person stage at age 6 is identical to the person stage at age 15, and that stage in turn is identical to the stage at age 35, which implies that the stages at ages 6 and 35 are identical as well).

---

[2] For more puzzles about identity of objects, see Nozick (1981).

[3] There is an alternative formulation of the problem of identity over time, based on which identity is conceived as the relationship that holds among continuant persons instead of among person stages. For more see, Lewis (1983).

When contemplating a normal course of an individual's life, it is clear that change occurs both physically and psychologically. Over time, most every cell in our body is replaced by a new one, and interests, preferences, memories, and other such psychological states modify a great deal as well. Yet a criterion of personal identity would have to establish that a person, for example, my neighbor Mary, remains one and the same across all of the stages of her life despite physical and psychological changes. To accommodate this problem, criteria for personal identity narrow the scope of properties to only those that are identified as necessary and sufficient to establish survival of a person over time. Hence, although Leibniz's law would not hold for the totality of an individual's features, there would be a subset of them that would remain unchanged over time, enough of those to say that the same person has survived over time.

One category of approaches selects particular physical features to establish identity over time, while another relies on the maintenance of psychological features. A traditional physical criterion establishes the relationship of identity between a person and her body, where the body excludes the brain (Perry 1978). This version of the bodily criterion excludes the brain because it defines identity in contrast to psychological criteria of identity, and the brain is assumed to be the seat of human psychology. This type of criterion is easily defeasible. A test of the intuitive acceptability of the bodily criterion is the following thought experiment: Imagine an accident in which two people, Jane and Mary, are injured. Jane's brain is destroyed in the accident but her body remains intact. After the accident, Mary's body is destroyed while her brain remains intact. Imagine further that Mary's healthy brain is transplanted into Jane's healthy body. Would it be accurate to say that Jane survived or that Mary is still alive? Intuitively, it would seem most accurate to claim that Mary survived the accident.

A more contemporary version of the physical criterion, and one impervious to the above criticism, is the biological criterion, which establishes identity between various stages of the same biological animal (DeGrazia 2005b). Based on this criterion, individuals are human animals persisting through the various stages of development of the body from birth through old age, including the various stages of brain development. According to this view, the body includes the brain. Based on the biological criterion, one can distinguish between the maintenance of personhood over time and the persistence of numerical identity. As David DeGrazia explains, "There was a time when we who are now persons were not persons (namely, before the human animal developed the capacities that constitute personhood), and there must be a time in the future when we are no longer persons (say, if severe dementia reduces us to barely sentient beings)" (Degrazia 2005b, p. 48).[4]

---

[4] Locke provides a similar biological criterion: "An animal is a living organized body; and consequently the same animal,…is the same continued life communicated to different particles of matter as they happen successively to be united to that organized living body. And whatever is talked of other definitions, ingenious observation puts it past doubt that the *idea* in our minds of which the sound *man* in our mouths is the sign, is nothing else but of an animal of such a certain form…" (Locke 1995 ed. 178).

The biological criterion does not capture what many have thought is what matters in identity across time, such as the maintenance of the same self, its experiences, memories, preferences, values, responsibilities, and so forth.[5] Psychological criteria, however, aim to establish the continuation of persons over time, and insofar as they require maintenance of identity of certain psychological features for survival, they qualify as numerical criteria of identity as well.

There are a number of distinct ways of formulating a psychological criterion. What they all have in common is that they allow in principle for the continuation of psychological features independently of the physical or bodily features of an individual. Locke (1995 ed.) formulates as follows: "personal identity consist: not in the identity of substance, but, ... in the identity of consciousness..." (p. 187).

Some psychological criteria identify the person with an immaterial soul. If there are such things as immaterial souls, then it is clear how they could survive the destruction of the body. But even if one has a more scientific conception of psychology, there is a way to argue that a person's psychology could survive the demise of the body. For example, if it were possible to exactly replicate the contents of Mary's brain, upload it into a fresh new one, one could argue that even the destruction of Mary's entire original body and brain would not be the death of Mary. What could signal a loss of personal identity by the psychological criterion are large changes in personality, including significant shifts in values, preferences, and long-term life plans (Shoemaker 1970, Perry 1972, Rorty 1976, Parfit 1984).

Some psychological criteria rely on the continuity of memories. Locke's criterion for "the sameness of a rational being," has been interpreted as a memory criterion for personal identity (Locke 1995 ed. 174–190). The reason to disagree with this interpretation is that Locke is not conditioning personal identity just on the persistence of certain aggregate memories, but on the *sameness of consciousness*. Locke argues that personal identity consists in the identity of consciousness, where "...it is the same self now it was then, and it is by the same self with this present one that now reflects on it, that that action was done" (pp. 180–181).[6] Locke's criterion might be better characterized as depending on a sense of continuity, rather than on the maintenance of particular memories.[7]

There are, however, memory criteria that identify the persistence of persons with the persistence of some core memories.[8] Such a criterion would of course have difficulty accounting for the maintenance of personal identity if a person were to become an amnesiac or forget most of his or her memories as a result of dementia.

---

[5] See, for example, Schechtman (1996).

[6] Locke notes that sameness of consciousness could be realized in a variety of distinct substances, but remains neutral as to the nature of the substance. But he does argue that personal identity depends on the existence of that substance. Thus, if the self is realized by Mary's pinky, and if the pinky becomes separated from Mary's body, she would survive as her pinky finger (Locke 1995 ed. 186).

[7] Whether a sense of continuity depends on the maintenance of particular memories will be discussed further in Chap. 5.

[8] Such a criterion is described in Perry (1978).

Finally, there are criteria that identify persons with the maintenance of core psychological traits, such as personality traits and values. Such a criterion is implicit in Parfit's example of the young Russian, as described in *Reasons and Persons*, who, motivated by his social ideals, decides to leave all of his wealth to his peasants after his death. Foreseeing a change in values as he grows older, the young Russian asks his wife to never let him revise his will (Parfit 1984, p. 327).[9]

However formulated, psychological criteria that require the maintenance of the relationship of identity over time run into difficulty reconciling the logical form of identity and the demand that a psychological criterion account for psychological continuity or as Locke puts it, the continuation of the same consciousness (Lewis 1983, p. 53).

The formal character of the identity requires that it be a one-to-one relationship. It cannot be the case that the car I drove to work yesterday is identical to both the car I drove to work today and the car I will drive to work tomorrow if the car I will drive to work today is a Honda and the one I will drive to work tomorrow is a Chevrolet. By analogy, it should also be that case that Mary at 6 years of age is identical to Mary at 15 and to Mary at 25 only if Mary at 15 and Mary at 25 are identical as well.

Psychological continuity defined as the continuation of certain psychological characteristics of an individual over time, the backbone of psychological criteria of identity, does not seem to have the same formal structure as the relation of identity. Psychological continuity can hold between one individual and many non-identical individuals (fission) and between many non-identical individuals and one individual (fusion). I will illustrate with a case of fission.

Consider the following thought experiment: Imagine that it is discovered that all that is needed for psychological continuity is the health of only one half of the human brain. An individual who has had half of her brain removed can be said to be psychologically continuous with the individual with an intact brain before surgery.[10] Imagine further that medical technology allows for half brains to be transplanted from one body to another, making it possible for a half brain from one individual to be transplanted into the body of another. For example, suppose that after a devastating car accident Mary is gravely injured and is dying. Mary's doctors decide to transplant half of Mary's brain into a different healthy body, that of Lucy. Lucy, who has also had a devastating car accident, has sustained injuries mostly to her brain but not to her body. Mary's half brain is transplanted into Lucy's healthy body. Because of the assumption that an individual with only half a brain can maintain psychological continuity, we can argue that psychological continuity will be maintained for Mary, but not for Lucy. Thus, we can say that Mary is the only one who survives the accident.

---

[9] For more on Parfit's example of the young Russian, see Chap. 6, Sect. 6.3.

[10] The justification of this premise is based on the conceivability of the claim that an individual who has lost half of her brain (due to a stroke, for example) could still be judged to be the same person as before she lost half of her brain (see Nozick 1981, pp. 39–40). The premise should not be construed as a factual claim about human psychology.

Suppose now that, to the doctors' surprise, Original Mary, whose body was badly damaged, continues living in the hospital for a period of time, say a month. Original Mary is psychologically continuous with Mary before the accident because, despite injuries, her body and the remaining half of her brain maintain her psychological continuity. During that same month, New Mary (Lucy's body plus one half of Mary's brain) recovers, leaves the hospital, and continues her life. New Mary is also psychologically continuous with Mary before the accident because she has half of her brain. Throughout the last month of Original Mary's life, there is a question about which one of the two Maries is identical with Mary before the car accident, Original Mary or New Mary.[11] Identity is a one-to-one relationship and only one of Original Mary and New Mary could be identical with Mary before the car accident. This case of fission, where one self purportedly divides into two, is illustrative of the presumed incompatibility between the formal character of identity and the concept of psychological continuity, indicating perhaps that psychological continuity is not best characterized by the relationship of identity.[12]

The incompatibility between identity and psychological continuity motivates Parfit to argue that personal identity is not what matters in survival (Parfit 1984, pp. 213–217). Parfit argues that in some cases, like the above case of fission, the answer about which Mary, Original or New, is identical to Mary before the accident would have to be solved arbitrarily. He formulates a different kind of relationship to undergird psychological continuity: the relationship of psychological connectedness and/or psychological continuity (relationship R). Relationship R does not have the formal character of identity and can hold between one individual and two others as in the fission case of Mary. Mary before the accident is psychologically continuous with both Original Mary and New Mary.

Psychological connectedness is achieved when the different stages of a person's life are connected by chains holding between past experiences and memories of the experiences, and the intentions and the acts in which those intentions are carried out (Parfit 1984, p. 206). Psychological connections are also present when a person continues to hold beliefs, desires, and ideals, and maintains a particular character or approach to life (Fields 1987). For instance, the person Mary is today bears a strong connection to the person she was yesterday, because there is a direct connection between the plans she made yesterday and the acts she is undertaking today. There

---

[11] The thought experiments described in this paragraph were adapted from a series of thought experiments presented by Nozick (1981, pp. 39–43).

[12] David Lewis (1983) argues that problems of fission can be resolved through a weaker relationship of identity, i.e., tensed identity. We speak of individuals being identical at a particular time. New Mary and Old Mary were identical at all the times before the car accident, but they are not identical after the accident. In this way we can preserve our manner of speaking of there being one Mary before the accident and two Maries after that, but without postulating the relationship of identity between the one Mary before and the two Maries after the accident (Lewis 1983, pp. 12–24).

is also a strong connection between Mary's actions yesterday and the memories she has of those experiences today.[13]

When there are enough strong connections between a person's past and future selves in terms of memories, goals, beliefs, desires, and intentions, there is psychological continuity between the two selves. Psychological continuity is achieved through overlapping chains of psychological connections. Psychological continuity is defined by reference to connectedness, where less connectedness entails diminished continuity of the self, and the relation of similarity will share this feature. The similarity relation is not transitive as is the relation of identity.

Let us consider an illustrative example: Mary at age 18 has several direct connections with Mary at age 12, since at age 18, Mary still has many clear memories of herself at 12, and she still has many of the same intentions and goals. At age 35, Mary has many vivid memories of her college years, when she was 18, but her attitudes and goals have changed significantly, and she barely remembers her adolescent years. Therefore, Mary at age 12 might be similar to Mary at age 18, and Mary at age 18 might be similar to Mary at age 35, but there will be less similarity between Mary at age 12 and age 35. The similarity criterion could accommodate degrees of similarity without disruption of continuity.

Abandoning the relationship of identity for that of similarity might come at a cost. Schechtman (1996) argues that we have certain pre-theoretical notions of personal identity, which accounts of psychological similarity do not satisfy (Schechtman 1996, pp. 51–66). For example, when I worry about surviving a turbulent airplane flight, I am not worried whether somebody sufficiently similar to me now will survive the flight; I am worried whether *I* will survive the flight. Identity, Schechtman (1996) claims, is required to explain why we are all particularly concerned about ourselves and in a way very different from how we are concerned about the states and interests of other people. Furthermore, moral responsibility and the concept of compensation require the utilization of personal identity. According to her, even accounts that preserve the concept of personal identity fail to capture the practical importance of identity for the sake of preserving logical consistency. In other words, none of the accounts thus far presented capture both the logical form of the identity relation and the practical importance of personal identity. Schechtman's response (1996) to this is to formulate an alternative account, that of narrative identity, which she argues captures the importance we accord to the notion of personal identity.

4.3.1 Narrative Identity

All the criteria thus far described could be categorized as third-person criteria of identity, distinct from personal conceptions of self. Even when devised to capture commonsense conceptions of identity, third-person criteria do not rely on features

---

[13] In this illustration, degrees of psychological connectedness trail temporal proximity, but that is not required based on Parfit's view.

of the self that are important to individual people. Rather, those devising such criteria select particular features of human psychology or biology that are the most likely to remain the same over time and could be used to establish the continuity of personal identity over time. To the extent that they are useful in resolving questions about how we remain one and the same person despite change, they apply to all people in the same way and in virtue of the same properties, either psychological ones or biological ones. While we might be able to establish a third-person criterion of identity without consulting actual people about their conceptions of self, we should not discuss narrative identity without relying on the features people deem important for their identity.

DeGrazia (2005b) describes our individual conceptions of self as constructions of narrative identities that can answer the question of "Who am I?" Each of us has an autobiography that provides an answer to that question. Schechtman (1996) calls this the self-constitution view, where personal identity over time is based on the ability of an individual to maintain a coherent and linear narrative about her self. Based on Schechtman's view, the ability to form a linear and coherent narrative is required for personhood, and individuals who are not capable of self-constitution in this way are not persons, although they might be owed moral regard nonetheless. Unlike numerical criteria of personal identity over time, Schechtman's self-constitution view aims to establish a relationship between the person and a particular trait, action, thought, or experience. Based on this type of account, the question becomes that of characterization, i.e., the degree to which a trait or an action could be said to characterize a particular person. A personal narrative should capture those traits and actions that characterize the person and capture the individual's true self. According to Schechtman, the self-characterization view captures our usual or commonsense notion of identity, one that is important when we wish to attribute praise or blame or explain why maintenance of identity is important. In addition, the self-characterization view concerns the sense of identity at issue when an individual is said to be undergoing an identity crises (Schechtman 1996, p. 74).

Assuming the narrative sense of identity, we can recast the objection to the use of cognitive enhancers as the worry about their causing individuals to suffer an identity crisis. This could happen when an individual experiences a change in personality that is difficult to incorporate into that individual's narrative. A person might experience a sense of discontinuity between the past and the present self and become uncertain about which self is the true self.

I argue that an identity crisis of this sort should be primarily conceived of as a first-person problem and whether an individual is experiencing such a break in personality can be only determined from that perspective. Consider the case of the half brain transplant described in the previous section. There, we argued that one could conceive of a psychological continuity account whereby psychological continuity is maintained even if an individual loses half of her brain to injury. This way of establishing psychological continuity might even cohere with the way in which we think of continuity of identity in actual cases where individuals suffer brain damage. Family members, friends, doctors, and others who know the person might speak of the person before injury and after injury as being the same individual. And insofar as they believe that there is a psychological continuant of the person before injury, they are likely to say that it is the person who survived the brain injury even if large parts of the brain are damaged.

If our focus is on the first-person sense of psychological continuity, it becomes important to determine also whether the individual thinks that there is psychological continuity before and after the injury whether she is able to incorporate whatever psychological and physical changes have occurred into a coherent narrative. Thinking back to Mary's fission case, one might wonder which one, Original Mary or New Mary, feels psychological continuity with the Mary before the car accident. It might be the case that one of the two, perhaps Original Mary, feels psychological continuity with the Mary before the accident. It might be that both feel continuous with the Mary before the accident or perhaps neither one of them feels psychological continuity. It might be the case that Original Mary is experiencing an identity crisis despite being psychologically continuous from the third-person perspective. It is possible then to have a discrepancy between the third-person perspective and the first-person perspective about whether there is psychological continuity. If the relevant sense of identity is the one we invoke when we think and speak about an identity crisis, then the first-person perspective should predominate.

The argument that cognitive enhancers might pose a threat to personal identity often refers to the possibility of the medication's changing individual personality traits, with those changes affecting an individual's conception of self. Moreover, such a change would be a change away from one's true self. My argument is that change of a particular personality trait need not lead to a disruption of a personal narrative or an identity crisis. Whether certain psychological changes actually cause an identity crisis or a disruption of narrative for an individual is an empirical question. I am not arguing that changes in traits never cause identity crises, but that there is no reason to assume that a change in traits will cause such a crisis. In the next section, I will present evidence about whether changes in traits result in disrupted narratives.

I will now sketch my view of narrative identity, which is closely aligned to the view endorsed by DeGrazia. My view is best evaluated keeping in mind several caveats. Although I endorse some of Schechtman's argument, I disagree with her that narrative identity provides an answer to the third-person problem of diachronic identity over time. Moreover, I do not link the ability to have a personal narrative to personhood, nor do I make strict requirements for what constitutes a narrative, as Schechtman does (1996, pp. 93–135). In addition, I think that the concept of narrative identity is separable from concepts such as autonomy, free will, or moral responsibility. Thus, my account should not be construed as providing any insights about those concepts. Finally, although my account of narrative identity can be construed as permissive, this should not be taken to imply that all actions that are in accord with one's true self are permissible.

Conceptions of self may depend on core values, beliefs, personal interests, and characteristics. It is important to note that the formation of a conception of self is based on the selection of certain traits. While there are many traits that might be true of an individual, not all of those can be said to characterize that individual. "A person is no more to be identified with everything that goes on in his mind,...than he is to be identified with everything that goes on in his body" (Frankfurt 1988, p. 61). There are things true of an individual of which she might be ignorant, such as the level of liver enzymes in her blood or the amount of potassium in her body. Furthermore, there might be psychological traits that are true of an individual, such as having poor long-term memory, that might not be important enough to be incorporated into that person's conception of self. A narrative requires a selection of certain traits that are important enough to become incorporated into a person's self-conception. Furthermore, a list of traits or characteristics true of an individual is not in itself a narrative. Rather, a person develops a narrative by incorporating some of her traits into the story of who she is. Thus, what is true of an individual can change over time without affecting narrative identity because identity is based on a selection of traits that are attributable to an individual.

There are individual differences in the constitution of self-conception. The traits I find most significant to my conception of self might not overlap with general characteristics that establish numerical identity over time. Although conceptions of self might superficially resemble psychological criteria for personal identity, they need not accomplish the same tasks as a numerical criterion of identity over time. When generating a conception of self, we need only to formulate a personal autobiography. The importance ascribed to features of the self is entirely individual. A person might choose her favorite characteristics and establish a hierarchy between them in any way she pleases. One might list her values, preferences, and personality traits and consider those features important to her narrative identity, even think of them as constitutive of her core self. Such personal impressions do not establish a more generally applicable criterion for personal identity, and one person's notion of personal identity does not restrict the conceptions of self for others. In other words, personal conceptions of self are not normative. They do not establish standards for what *should* be true of conceptions of self.

Each of us could prioritize the elements of our core self in very different ways. Some might prioritize their professional identity over any other kind of interest, value, or personality feature. Others might define themselves in terms of the role they have in their family, as daughter, mother, or favorite aunt. One might have inaccurate conceptions of one's self. For example, a person might think of herself as having a great sense of humor, and perhaps make elaborate efforts to be humorous, even if such an impression is not shared by others. In fact, there could be significant differences between how we think of ourselves and how others see us.

Conceptions of self need not capture what are often thought of as important features of personal identity. A person's core identity need not include traits that others consider constitutive, such as religion, gender, sexual orientation, nationality, profession, and so on. One could prioritize relatively unimportant personal features, such as being thrifty or being a good dresser, as part of one's core identity and make efforts to act in accordance with that self-image. As we change over time, grow older, or learn new things, the relative importance of our core traits might change. These changes in identity can be traced back to a variety of sources. A person may change her social status, obtain a college degree, become part of a family unit, or even join a religious cult.

A criticism of my very liberal conception of narrative identity is the requirement of accuracy. Schechtman (1996) argues that our narrative identity should cohere with what others think of us (p. 95). She also distinguishes between errors of fact and errors of interpretation, where errors of interpretation are more acceptable than errors of fact (pp. 121–130). If an individual's narrative identity is based on factual errors, that person has not characterized herself properly. For example, an individual who is not Napoleon might adopt Napoleon's narrative and stake a claim to his identity. In such a circumstance, a good account of narrative identity should be able to explain why it would be wrong for an individual who is not Napoleon to adopt his narrative.

I think that Schechtman is correct in arguing that some narratives could be deemed inadequate when they are replete with factual errors and that there ought to be a requirement that a good narrative should not be based on factual errors. This requirement, however, is very limited in scope and will rule out only very few narratives as inadequate. The accuracy of much of what is included in a narrative is based on evaluation. For example, I might not think that I am Napoleon, but I might think that I am as courageous as he was. Furthermore, my family and I might disagree about whether I am courageous. They might think me quite timid, while I might present the story of my life as a tale of one courageous feat after another.

In order to argue that I am wrong about myself and that my narrative fails to characterize me accurately, one would have to maintain that it is other people, my friends and family, who are able to know more accurately what is true of me than I can. This strikes me as false. My claim is not based on the notion that each individual has privileged access to who she is; rather, it is an argument that what we think of others and of ourselves is not often based on facts, but on interpretation. Requiring coherence, then, with how others see us will not increase the accuracy of personal narratives.

There could be a number of reasons for discrepancies between how I characterize myself and how others do. For example, my family might think I am timid because they base their judgment on limited evidence. They see me only on very few occasions, such as family gatherings, where I tend to be timid. We might have different interpretations of courage. As an introvert, I might think that talking to strangers is courageous, while my family might think that courage only applies to cases where an individual exposes herself to bodily harm. Their interpretation of me might be influenced by how they characterize themselves. For example, my siblings might wish to think of themselves as courageous and they do so by comparing themselves with me. They interpret me as timid in order to feel courageous in comparison. Finally, being around my family, where everybody thinks of me as a wimp, might suppress my disposition to be courageous, in effect concealing my true self.

Furthermore, as individuals change over time, some traits that could be attributable to them might become less prominent and others might take their place. Over time a person who is shy might become less so and ultimately become unabashed. Such a change might occur gradually and the individual might have periods where she acts both shy and confident, making the two traits equally attributable to her. Thus, if one were to wonder whether a particular trait is attributable to Jane, it might be difficult to determine which one of the traits, either shyness or confidence, is part of her core self. She might select either one of those two characteristics as true of herself, or she might incorporate them both into her narrative, admitting that she is sometimes shy and sometimes confident, depending on the situation.[14] In such cases, it is the personal endorsement of a particular trait that makes either shyness or confidence true of one's self.

Given that the extent to which particular traits are attributable to certain individuals is a matter of interpretation—either the interpretation of the individual attempting to form a narrative self or that of those who know that individual—the requirement that individual self-conceptions should coincide with what others think of that individual is not justified. This is not because there are no means of distinguishing between evaluations in terms of quality, but because there is no reason to think that someone else's evaluation of an individual is better than the evaluation of the individual herself as they pertain to narrative identity. As I have argued just a few paragraphs ago, the interpretation of other individuals as to whether a certain trait characterizes an individual can be plagued by biases, or limited or selective evidence, as can the individual's own interpretation. Furthermore, whether a narrative accurately reflects an individual's true self can be challenging to establish as we change over time. There might be periods when it is indeterminate if a certain trait can be properly attributed to a person and the character of one's true self would be fixed only by personal endorsement.

4.3.2 Change and Disruption of Narrative Identity

Even if cognitive enhancers might precipitate a change in personality, a person's self-characterization might change to accommodate new personality traits. Research by Quoidbach, Gilbert, and Wilson (2013) supports the claim that people change a great deal over time. The study found that changes occur both in personality traits, as in expressed commitment to particular values, and as they occur in personal preferences. All those features seem to correspond neatly with common elements used to define the notion of narrative identity. Interestingly, the study found that when asked to predict how much they will change in the future, participants tended to underestimate the amount of future change to their narrative identity. The participants in the study were able to assess accurately the amount of change they underwent in the past, but kept their conviction that their current cluster of personality traits, values, and preferences would persist in the future. It seems then that a concept of self can be maintained despite change.

Given that continuity can be maintained despite change to even purportedly important features of the self, it seems wrong to argue that cognitive enhancers

---

[14] For a further argument on the indeterminacy of the true self, see Chap. 5, Sect. 5.3.

should not be used because they might change core personality traits. If individual traits can vary as a result of education, change of religion, or shifts in social and other circumstances, then alterations in narrative identity that result from cognitive enhancement ought to be conceived of as just one kind of potential change to self over time. Presumably the objection to change that results from cognitive enhancement is that it is not naturally occurring. The distinction between natural and unnatural changes in general strikes me as particularly difficult, if not impossible, to establish. But even without challenging that distinction, when one accepts that some choices made by individuals that might result in a changed self—such as joining a yoga class or going to a therapist—are morally permissible, then it is arbitrary to argue that the use of neurocognitive enhancers is impermissible because it might result in a changed self. Furthermore, there are instances where enhancers could be seen as promoting continuity of self, by restoring levels of cognitive performance to where they were before the deleterious effects of sleep deprivation, illness, or fatigue. The distinction between permissible and impermissible forms of enhancement here seems to rely merely on tradition, which could change as pharmaceutical means of promoting authenticity become more prevalent and join the ranks of established ways to change and improve the self.

Returning to the issue of identity crisis, an argument against the use of enhancers could be that they could cause psychological change that fails to incorporate into a personal narrative and thereby render it incoherent. This could happen if taking medication could suddenly precipitate a change in a relevant personality trait that would outstrip the individual's ability to accommodate such changes into a personal narrative. Limited evidence indicates that persons do not experience the effects of enhancers the same way. Some feel more and others less like themselves after taking medication. In a study done by Bolt and Schemer (2009), patients treated for ADHD reported a variety of perspectives when it came to the influence of medication on their sense of personal identity. Although some study participants said that taking medication felt like it changed their personality, some reported that they felt more like themselves (Bolt and Schermer 2009). Kramer (1997) reports something similar in his description of a patient taking medication for depression; the patient reported that she felt more like herself on medication than when she was not taking it.

This conflicting evidence about the perception of the continuity of self shows that one cannot conclude that the use of neurocognitive enhancers will result in a changed identity just because it might effect changes in some personality traits. Moreover, it is not true that cognitive enhancers, as described in Sect. 4.2, caused unusually notable change in cognitive function, only moderate improvements. Based on the evidence presented in Sect. 4.2, it is not possible to know whether the change in performance, where there was any, was noticed by the participants in the studies.

Setting the issue of empirical evidence aside, it is not obvious that sudden changes in personality traits, even those that were central to an individual's narrative, would cause an identity crisis. Sudden changes in personality might be brought on very quickly even without the use of enhancers. One could experience an impactful event that might cause fundamental changes in personality. An individual might suddenly become apprised of a fact that affects her priorities in some important way. For example, one could visit an orphanage, become aware of injustice and economic disparity, and become charitable. In some cases, even if the change in trait is sudden, it might become incorporated gradually as the person slowly becomes aware that she has become different in a certain way. This would then abate the impact of sudden change because it would be experienced gradually.

A different argument against the use of enhancers could be that they could disrupt the causal chain maintained between our psychological states over time. Parfit (1984) and Nozick (1981) argue that psychological continuity is maintained when current person stages are causally dependent in the right way on previous ones. The argument against enhancers could be that they effect personality changes, but do not preserve the right type of causal dependence on the individual's previous person stages and are thus likely to cause psychological discontinuity.[15] Again, whether the use of enhancers could result in a felt psychological discontinuity is an empirical question. But it is possible to argue that such an effect should not be presumed.

Taking for granted the least exigent physicalism, which is only that our psychological states are somehow causally dependent on physical states, one can argue that all of our person stages are caused by some physical process in the brain. As the brain changes or is affected by hormonal and other relevant influences, those changes can influence psychology. For example, changes in levels of neurotransmitters epinephrine and norepinephrine, which are sometimes activated in stressful situations, will affect the quality of our memory.[16] In traumatic situations there is an increase in the production of epinephrine and norepinephrine, which can have the effect of enhancing memory of the traumatic event (Cahill and McGaugh 1996). The quality and vividness of those memories can influence future person stages by affecting the ability to cope with a traumatic event. Such an event might even produce a change in narrative identity, as significant events sometimes have that effect. Medicines called beta-blockers can help decrease the secretion of epinephrine and norepinephrine, which in turn prevents the enhanced memory of the traumatic event (Reist et al. 2001). Using beta-blockers can abate the influence of the traumatic memory on future person stages, thereby perhaps preventing a change in narrative identity.

Since changes in the brain are a requirement for psychological change, person stages could be said to be dependent on the physical stages of the brain. Moreover, the psychological changes resulting from a traumatic event or the prevention of those changes through the use of beta-blockers exploit a similar physical mechanism, making it difficult to argue that one type of change, increased level of hormones, preserves causal dependence among person stages, while the other type of change, decrease in hormonal levels, does not. Hence, if the argument is that the use of enhancers is impermissible because it can affect our psychology by affecting the brain, then the argument does not succeed. Psychological changes are the outcome

---

[15] I am not attributing this view to either Parfit or Nozick. In fact, Parfit adopts a very permissive view on what could constitute "the right kind of cause" (Parfit 1984, p. 215).

[16] For a complete discussion of this phenomenon, see Chap. 5, Sect. 5.2.

of physical changes in the brain, but psychological discontinuity need not be the outcome of changes in brain function. In fact, taking beta-blockers could be said to help preserve narrative identity and psychological continuity.

A slightly different argument against enhancers could be constructed in the following manner: One could argue that there are two parallel streams of causal chains, the causal chain among the physical processes in the brain and the causal chain among the supervenient psychological states. The psychological chain would be causally influenced by the physical causal chain, but the manner in which the psychological states cause each other at the same level would be different from the type of causation that exists at the physical level. The causal chain at the level of psychological states is maintained in terms of the properties of those states, such as their content. For example: I remember that a few years ago I visited an orphanage and that the children there were living in squalid conditions. This memory in turn prompted me to believe that such living conditions were not morally acceptable, which in turn prompts me to donate money to the orphanage. The claim would be that use of cognitive enhancers would disrupt this type of causal chain, where mental states cause each other in terms of their content. My argument previously was that psychological continuity would be maintained even with the use of cognitive enhancers because they exploit brain processes in order to effect change in psychology in ways that are not relevantly different from other mechanisms that result in psychological change. But that argument leaves open the possibility that there could be a disruption in the causal chain because enhancers would have changed the content of my beliefs in a way that was not the result of the states earlier in the causal chain.

The type of causal chain I described that led me to donate money could be the result of any number of physical processes, none of which I could be able to distinguish from the first-person perspective. My having a thought process of the type described above is compatible with my being a brain in a vat or having hydraulics for a brain, like Lewis's Martian in pain (Lewis 1991 ed.). Thus, a change in the physical processes that underlay my psychological continuity might not be noted from the first-person perspective. If I were to take enhancers, I might notice a change in my tendencies. For example, when remembering the appalling conditions in the orphanage, I might no longer be motivated to think that such conditions are unacceptable. I might instead have the thought that disparity is the way of the world and then fail to be motivated to write a check to donate money. In both cases, the one where I am prompted not to accept inequality and the one where I do, the memory is the same. The way in which it causes or fails to cause a thought about the acceptability of socioeconomic disparity is in terms of the content of my memory and the content of other occurrent beliefs. Hence, the causal chain is maintained, despite my taking medication, and despite my reacting differently than before. The requirement that causal dependence at the psychological level be maintained cannot be that the contents of psychological states remain the same even in similar situations (e.g., in relevant situations always thinking that economic disparity is morally unacceptable), just that they cause each other in terms of their contents, whatever those might be.

Finally, opponents of the use of enhancers could revert back to the argument that attempts to draw a moral distinction between the causes of psychological change. Based on their view, it might be permissible to change as a result of going to school or becoming more religious, but it is not permissible to endure psychological changes as a result of cognitive enhancement. This argument seems even less supported than before. We know that changes occur over time for most individuals, and that those changes are of different psychological aspects. We further know that in order for any influence on our psychology to be effected, it needs to exploit some type of physical process in the brain. Moreover, sometimes those processes are similar, as we have seen in the example of the traumatic event and beta-blockers. Thus, it becomes arbitrary to argue that willful change of self is acceptable unless it comes as a result of the use of medication.

4.4   Conclusion

In this chapter, I evaluate the arguments that cognitive enhancement could cause disruptive changes in personal identity. Arguments that the use of enhancers could disrupt personal identity are often based on the confusion between numerical and narrative identity. I distinguish between those concepts. I argue that cognitive enhancement would have no effect on numerical identity; rather, it could affect only narrative identity. Narrative identity is a first-person effort to construct a concept of self based on some, out of many, traits that could be attributable to an individual. Particular traits can change without altering the narrative self and even when psychological changes do become incorporated into a personal narrative, they can occur without disrupting the sense of self over time. There is evidence that individuals change a great deal over time, as there is evidence that even when those changes are caused by the use of medication, they do not negatively affect narrative identity. Accordingly, I conclude that cognitive enhancement is permissible even when it effects changes in psychology.

References

Advokat, C. D. (2010). What are the cognitive effects of stimulant medications? Emphasis on adults with attention-deficit/hyperactivity disorder (ADHD). *Neuroscience and Biobehavioral Reviews, 34*, 1256–1266.

Advokat, C. D., Guidry, D., & Martino, L. (2008). Licit and illicit use of medications for attention-deficit hyperactivity disorder in undergraduate college students. *Journal of American College Health, 56*(6), 601–606.

Beglinger, L. J., Gaydos, B. L., Kareken, D. A., Tangphao-Daniels, O., Siemers, E. R., & Mohs, R. C. (2004). Neuropsychological test performance in healthy volunteers before and after donepezil administration. *Journal of Psychopharmacology, 18*, 102–108.

Bolt, I., & Schermer, M. (2009). Psychopharmaceutical enhancers: Enhancing identity? *Neuroethics, 2*, 103–111.

Cahill, L., & McGaugh, J. L. (1996). Modulation of memory storage. *Current Opinions in Neurobiology, 6*, 237–242.

Chatterjee, A. (2004). Cosmetic neurology: The controversy over enhancing movement, mentation, and mood. *Neurology, 63*, 968–974.

Daniels, N. (2000). Normal functioning and the treatment–enhancement distinction. *Cambridge Quarterly of Healthcare Ethics, 9*, 309–322.

DeGrazia, D. (2005a). Enhancement technologies and human identity. *Journal of Medicine and Philosophy, 30*, 261–283.

DeGrazia, D. (2005b). *Human identity and bioethics*. New York: Cambridge University Press.

Elliot, C. (1999). *A philosophical disease: Bioethics, culture and identity*. New York: Routledge.

Elliot, R., Sahakian, B. J., Matthews, K., Bannerjea, A., Rimmer, J., & Robbins, T. W. (1997). Effects of methylphenidate on spatial working memory and planning in healthy young adults. *Psychopharmacology, 131*, 196–206.

Farah, M. J., Illes, J., Cook-Deegan, R., Gardner, H., Kandel, E., King, P., et al. (2004). Neurocognitive enhancement: What can we do and what should we do? *Nature Reviews, 5*, 421–425.

Fields, L. (1987). Parfit on personal identity and desert. *The Philosophical Quarterly, 37*, 432–441.

Frankfurt, H. (1988). *The importance of what we care about*. New York: Cambridge University Press.

Glannon, W. (2007). *Bioethics and the brain*. New York: Oxford University Press.

Gligorov, N. (2010). Seeking more than health: Using medicine for enhancement. *American Academy of Pediatrics Newsletter—Section in Bioethics, 15*–18.

Grady, S., Aeschbach, D., Wright, K. P., Jr., & Czeisler, C. A. (2010). Effect of modafinil on impairments in neurobehavioral performance and learning associated with extended wakefulness and circadian misalignment. *Neuropsychopharmacology, 35*, 1910–1920.

Hall, K. M., Irwin, M. M., Bowman, K. A., Frankenberger, W., & Jewett, D. C. (2005). Illicit use of prescribed stimulant medication among college students. *Journal of American College Health, 53*(4), 167–174.

Izquierdo, I., Bevilaqua, L. R., Rossato, J. I., Lima, R. H., Medina, J. H., & Cammarota, M. (2008). Age-dependent and age-independent human memory persistence is enhanced by delayed post-training methylphenidate administration. *Proceedings of the National Academy of Sciences, 105*, 19504–19507.

Kramer, P. D. (1997). *Listening to prozac: A psychiatrist explores antidepressant drugs and the remaking of self*. New York: Penguin.

Lewis, D. (1980, 1991 ed.). Mad pain and Martian pain. In D. Rosenthal (Ed.), *The nature of mind* (pp. 229–233). New York: Oxford University Press.

Lewis, D. (1983). Survival and identity. In *Philosophical papers I* (pp. 55–77). New York: Oxford University Press.

Locke, J. (1690, 1995 ed.). *An essay concerning human understanding* (pp. 174–190). London: Everyman Press.

Mehta, M. A., Owen, A. M., Sahakian, B. J., Mavaddat, N., Pickard, J. D., & Robbins, T. W. (2000). Methylphenidate enhances working memory by modulating discrete frontal and parietal lobe regions in the human brain. *Journal of Neuroscience, 20*(6), RC65.

Nozick, R. (1981). Personal identity through time. In *Philosophical explanations* (pp. 29–71). Cambridge, MA: The Belknap Press.

Parfit, D. (1984). *Reasons and persons*. Oxford: Clarendon.

Perry, J. (1972). Can the self divide? *Journal of Philosophy, 69*(16), 463–488.

Perry, J. (1978). *A dialogue on personal identity and immortality*. Indianapolis: Hackett Publishing Company.

Presidential Council on Bioethics. (2003). *Beyond therapy: Biotechnology and the pursuit of happiness*. Washington, DC: PCB.

Quoidbach, J., Gilbert, D. T., & Wilson, T. D. (2013). The end of history illusion. *Science, 339*, 96–98.

Reist, C., Duffy, J. G., Fujimoto, K., & Cahill, L. (2001). Beta-adrenergic blockade and emotional memory in PTSD. *International Journal of Neuropsychopharmacology, 4*, 377–383.

Repantis, D., Laisney, O., & Heuser, I. (2010). Acetylcholinesterase inhibitors and memantine for neuroenhancement in healthy individuals: A systematic review. *Pharmacological Research, 61*, 473–481.

Rorty, A. (1976). *The identities of persons*. Berkeley: University of California Press.

Sandel, M. (2004). The case against perfection. *The Atlantic Monthly, 293*(3), 51–62.

Schechtman, M. (1996). *The constitution of selves*. Ithaca: Cornell University Press.

Shoemaker, S. (1970). Persons and their past. *American Philosophical Quarterly, 7*, 269–285.

Sugden, C., Housden, C. R., Aggarwal, R., Sahakian, B. J., & Darzi, A. (2012). Effect of pharmacological enhancement on the cognitive and clinical psychomotor performance of sleep-deprived doctors. *Annals of Surgery, 255*(2), 222–227.

Yesavage, J. A., Mumenthaler, M. S., Taylor, J. L., Friedman, L., O'Hara, R., Sheikh, J., et al. (2002). Donepezil and flight simulator performance: Effects on retention of complex skills. *Neurology, 59*, 123–125.

Chapter 5
The Truth About Memory and Identity

Abstract The moral condemnation of memory modifying technologies (MMTs) often relies on the view that memory provides a veridical representation of the past and that it can be used to ground personal identity. In this chapter, I present a range of studies that substantiate the claim that autobiographical memory is unreliable and cannot be used to ground narrative identity. I use this evidence to argue that MMTs that have the potential to alter autobiographical memory do not jeopardize personal identity. Given its flexibility, I argue, narrative identity can be maintained despite changes in memory. I further argue that maintenance of particular memories is not required for authenticity. Because of the spontaneous fluctuations of each person's character traits, values, and preferences over time, I claim that first-person endorsement of core traits or the identification of core memories as formative of narrative identity is required to establish one's true self. In addition, I dispute the argument that memory modification poses a challenge to authenticity and provide examples of instances where such modification can promote authenticity.

5.1 Introduction

Research on human memory has revealed the precarious nature of remembrance. Memories are not a veridical representation of the past. In fact, memories appear to be reconstructed at the time of recall as they are affected by events that immediately follow the events remembered, such as conversations, or manipulative or probing questioning (Loftus 2003).

There are additional ways of shaping memory. Liao and Sandberg (2008) coin a term, memory modifying technologies (MMTs), to designate the different ways in which memory can be altered, using either medication or suggestion. The two directions of such alterations are either to eliminate or diminish some memories or to improve memory overall. MMTs have been discussed in terms of their ethical impact, similar to those for cognitive enhancements, i.e., altering an individual's memory might lead to changes in personal identity. Either diminishing or improving memories could affect personal identity by either erasing crucial memories or by expanding the assemblage of memories that might become formative of identity. An example of an MMT is propranolol, a beta-blocker commonly used to treat hypertension, tremor, and migraines, which has also been shown to be efficacious as a

© Springer Science+Business Media B.V. Dordrecht 2016   75
N. Gligorov, *Neuroethics and the Scientific Revision of Common Sense*,
Studies in Brain and Mind 11, DOI 10.1007/978-94-024-0965-9_5

memory modifier for traumatic events. The potential use of propranolol as a memory modifier has produced a range of responses (Kass 2003; Liao and Sandberg 2008; Henry et al. 2007; Wasserman 2004). Some of those include concerns about whether diminishing the memory of formative events, even those that are negative or traumatic, could disrupt a person's sense of authenticity (Erler 2011).

In previous chapters, I utilized the distinction between numerical identity and narrative identity. I argued that cognitive enhancement will not affect our sense of narrative identity. In this chapter, I extend that argument to the use of MMTs. In Sect. 5.2, I review research on the ways in which memory is not reliable and the ways in which learning new information can affect our memory of the past. I will also discuss some studies showing memory insertion. I use this research as the basis for my argument that memory is not a dependable source of self-knowledge.[1] In addition, I review some of the currently available MMTs. In Sect. 5.3, I evaluate the impact of changes in memory on personal identity. Because of the unreliability of memory, I argue that autobiographical memory cannot ground either our knowledge of the past or our knowledge of our previous stages of self. Furthermore, I argue that, in order to maintain that particular memories are relevant to personal identity, one must revert back to using a numerical criterion of identity.

In Sect. 5.3 of this chapter, I evaluate the charge that MMTs could pose a threat to one's true self. Specifically, I evaluate Alexander Erler's criticism of what he calls the existentialist view of authenticity (Erler 2011). Erler proposes that a person's endorsement of certain personal traits is not sufficient to circumscribe an individual's true self and proposes instead that one's true self has to be based in part on others' impressions of one's core self. I argue against this view, challenging the claim that traits of one's core self can be established in nonsubjective ways. Here also, I argue that in order to establish a person's core self without relying on that individual's own endorsement of such traits, one has to rely on a numerical criterion of personal identity.

The upshot of my argument is that because memory modification does not disrupt narrative identity, the use of them is permissible. My argument, however, leaves open the possibility that instances of memory modification could be impermissible, for example, in cases where the negative side effects to an individual outweigh the benefits.

5.2 Interfering with Memory

Retroactive interference with memory has been widely studied as a potential cause of forgetting. In 1900, Muller and Pilzeker conducted an experiment for which they divided participants into two groups. One group was given a list of words to memorize, while the other group was given that same list of words and then, shortly

---

[1] My review of the scientific literature on memory is limited by my goal of assessing the argument that memory modifying technology could disrupt narrative identity.

after, one more list of words. The groups were then tested on their recall of the original list. What they found was that the group that had to memorize a second list had significantly worse recall of the first list. This experiment demonstrated the effect of retroactive interference, where the learning of new information following an event interferes with the memory of that event. Subsequently, there were a number of other experiments that confirmed the effects of retroactive interference.[2]

Elizabeth Loftus has investigated the effects of retroactive interference for witnessed events. Loftus (2003) argues that memories are not fixed and that in addition to being lost they can be changed over time. The process of recall is not rightly characterized as a mere accessing of past experience; rather the act of recall is an act of reconstruction. Loftus and Palmer (1974) showed that the way in which people are questioned about a witnessed event can influence the way they remember it. For example, in the study, participants were shown a film about a car accident. Participants were asked to answer questions regarding the details of the accident. One group of participants was asked: "How fast were the cars going when they *smashed* each other?" while the other group was asked: "How fast were the cars going when they *hit* each other?" The estimated speed changed based on the question. Those who received the questions with the word 'smashed' in it, judged the speed of the cars to be significantly faster than those who received the question with the word 'hit.'

This experiment further showed that participants combined the memory of the event with the memory of the subsequent questioning into one. After being asked about the speed at which the cars were going, the participants were also asked to remember the amount of broken glass that resulted from the car collision. They were asked, "Did you see any broken glass?" Those participants who were asked about the cars *smashing* into each other said that they recalled there being a lot more broken glass than those participants who were asked about the cars' *hitting* each other. In fact, there was no shattered glass in the film about the car accident. Loftus and Palmer argued that the misinformation about the event actually changed the original memory of the event and caused those who were asked about the smashing of the two cars to reinterpret the original event as more serious (Loftus and Palmer 1974).

A subsequent study by Loftus (1977) further supports retroactive interference as a cause of forgetting. In the study, the participants watched a videotaped simulated crime or accident in order to test their recall of these simulated events. After watching these videos, the participants in the study were given false information about the event in which the details of the crime or accident scene were changed. The misinformation was subtle. For example, the participants were asked the following question about the viewed car accident: "Did the blue car that drove past the accident have a ski rack on the roof?" The car in the video had in fact been green. The participants would then be asked to recall the details of the accident, which usually involved a forced choice, say between a green and a blue car. Those participants who were exposed to the misleading information were significantly more likely to

---

[2] For additional studies on retroactive interference, see Green (1992), pp. 154–156.

choose 'blue' as the correct answer. It is of note that the misled participants have as much confidence in their judgments as those who were not given any misleading information and were more likely to be accurate in their judgments (Loftus et al. 1989).

Loftus's proposed the *substitution hypothesis* as an explanation for her results. According to this hypothesis, new memories can override established memories; the new information causes the erasure of the old information. It is not just that we are no longer able to recall the information; it is that the new, and sometimes misleading, information, actually becomes our memory of the event. There is an alternative explanation for the data, which is the biased-guessing interpretation put forth by McCloskey and Zarazoga (1985). They argue that the misleading information does not actually replace the initial memory of the event; rather, it creates a response competition between the two remembered alternatives, one for the witnessed event and one reflecting the misleading information. Therefore, when the participants were asked to complete a forced-choice task between the details as they happened in the video and as they were presented in the misleading information, the misleading information biases recall in favor of the misleading alternative. Based on this account, the original trace of the witnessed event was not extinguished by the new memory for the misleading information, but the misleading information produced an error during the process of remembering. Mcloskey and Zaragoza changed the paradigm used by Loftus to a choice between the original stimulus and a novel stimulus, thereby circumventing response competition (Green 1992). Experiments using this modified methodology found no difference between groups that were giving misleading information after the witnessed event and those who were not misled. Nonetheless, either because of substitution or response competition, both these paradigms indicate that memory is not a reliable record of past events.

In further experiments, Loftus (2003) showed that it is possible to influence people to remember events that never happened to them. In her study, Loftus gave participants a booklet that contained descriptions of four events from the person's past. Three of the four experiences were events that actually happened, and the details of those events were gathered from the individual's family members. The fourth event, which described an episode of being lost in the mall, was fabricated. The results of the study showed that 29 % of study participants reported remembering the memory of being lost in the mall although the event did not actually happen to them. This experiment purports to show the successful insertion of false memories. In another experiment with memory insertion, participants were shown a doctored photograph, where a real photograph of the participant was inserted into a photograph of an air-balloon ride (Wade et al. 2002). For the sake of the study, family members confirmed that the balloon ride happened. The participants were then asked to remember everything they could about the balloon ride. About half the participants in the study recalled either partially or completely having been on the balloon ride.

For further evidence of the unreliability of memory, a study by Sheen et al. (2001) challenges even the reliability of the feeling of ownership of our memories. The study recruited twin participants who were asked to generate a list of autobiographical memories in response to cue words, including bicycle, birthday, and holiday. In the course of the interview with the twin participants, a number of memories were described whose ownership then became disputed; both twins claimed that it happened to them. The twins agreed about the details of the experience, and both reported on the vividness of the memory by elaborating on the visual details of the event. In a subsequent study, Sheen et al. (2006) analyzed the disputed memories, most of which were childhood memories that fell into a number of categories, such as memories of wrongdoing, misfortune, receiving a gift, daring. Predictably, subjects were more likely to report ownership of memories if they involved misfortune and achievement and less likely to claim ownership if the memory was of wrongdoing.

The studies by Sheen et al. (2001, 2006) and the research by Loftus document the natural unreliability of memory. But memory can be influenced in more interventional ways, through the use of pharmaceutical agents. There are a number of drugs that can enhance memory in normal individuals. In Chap. 4, we discussed stimulants, such as methylphenidate (Ritalin®) and dextroamphetamine (Adderall®), which have been shown to improve spatial working memory, working memory, and long-term memory, if taken shortly before the event to be recalled. In a study by Mehta et al. (2000), methylphenidate was shown to produce improvement in working memory. Also, in a study by Elliot et al. (1997) methylphenidate was shown to improve spatial working memory on some tasks, although the improvements were seen only when the task was novel. Finally, in a study by Izquierdo et al. (2008) methylphenidate was shown to improve long-term memory in individuals who were over the age of 35.

There are also pharmaceutical agents that have been shown to be memory disrupters; most specifically, they interfere with the formation of long-term memories. Scopolamine has been shown to affect the initial stages of memory acquisition, such as encoding and consolidation as well as spontaneous memory retrieval (Caine et al. 1981). Scopolamine had no effect on retention of information (Caine et al. 1981). Benzodiazepines have been shown to cause temporary anterograde amnesia—the inability to form new memories. The dose of the drug, rate of absorption, and method of administration modulate the effects of benzodiazepines on memory (King 1992).

Finally, beta-blockers, such as propranolol have been shown to affect memory retention. Emotional arousal has been documented to have an enhancing effect on memory (Cahill and McGaugh 1996). Propranolol is thought to block the release of the stress hormones epinephrine and norepinephrine, which, together with the activation in the amygdala, have been shown to form the biological basis of our emotional responses to traumatic events (Reist et al. 2001). In a placebo-controlled trial by Cahill et al (1994), subjects viewed either an emotional story or a neutral story. Some of the subjects in the study were given propranolol one hour prior to viewing the stories, and all of the participants were tested for recall 1 week after the viewing. The recall of the arousing story was better than the recall for the neutral story in the placebo group. For those participants who took propranolol before viewing the story, the recall of the story that depicted a traumatic event was diminished. Propranolol had no effect on the recall of the neutral story.

A further study by Reist et al. (2001) confirmed these results. The study compared army veterans with post-traumatic stress disorder (PTSD) and normal controls. Reist et al. (2001) used the same methodology as Cahill et al. (1994). PTSD is a disorder of memory with intrusive recollection of past traumatic events. It is characterized by recurrent daytime memories, nightmares, and flashbacks. The traumatic memories can be triggered by a variety of everyday stimuli (Reist et al. 2001). The participants in the Reist et al. (2001) study viewed an arousing and a neutral story. Some of the participants were given propranolol prior to viewing the story, and all the subjects were tested for recall. For those participants who did not take propranolol, the arousing story enhanced recall, and as expected, propranolol blocked that effect for the other participants. There were no differences in recall between the control group and the participants with the diagnosis of PTSD. Moreover, propranolol affected both study groups in the same way.

Because of the effects of propranolol on arousal and the corollary impact on memory of traumatic effects, Pitman conducted a study to investigate the efficacy of beta-blockers in preventing PTSD. Pitman postulated that the excess of epinephrine at the time of the traumatic event leads to an overly strong emotional memory and fear conditioning that subsequently results in PTSD (Pitman 1989). To test this hypothesis, Pitman et al. (2002) conducted a pilot study in an emergency department (ED) on subjects who had just experienced a traumatic event, as defined by the Diagnostic and Statistical Manual of Mental Disorders (DSM) IV (American Psychiatric Association 2000). The subjects were randomized to a placebo and a propranolol arm of the study. Those who were in the propranolol study were administered a dose of propranolol in the ED and then instructed to continue taking that medication for another 10 days. The rate of PTSD in the control group was 30 %, and in the propranolol group it was 18 %. The difference between the two groups was statistically significant. The results of the study indicate that propranolol might have preventive effects if administrated after the traumatic event.

In another study, Vaiva et al. (2003) applied a similar methodology as in Pitman et al. (2002), where propranolol was administered after a traumatic event for 7 days, with a taper period of 8–12 days. This study, however, was not randomized. Participants who were unwilling to take propranolol, but were willing to participate in the study were also enrolled. Vaiva et al. (2003) then compared the rates of PTSD in the group of participants willing to take propranolol and in those who refused the medication. The study found a reduction in the rate of PTSD for those individuals who agreed to take propranolol. Although the study was not randomized, it did not significantly differ in demographic, severity of injury, or emotional responses from those enrolled in the Pitman et al. (2002) study.

It is of note, however, that in a most recent randomized control study by Hoge et al. (2012), propranolol did not have any protective effects against PTSD. This study recruited participants from the ED for 4 years from 2004 to 2008. The recruited participants were both men and women between the ages of 16 and 65 who had just experienced a traumatic event, as defined by the DSM-IV. In this study, the participants were randomized into a placebo or a propranolol arm of the study. Those who were randomized into the propranolol arm were given a dose of propranolol upon arriving to the ED and were asked to continue taking it for a total of 19 days. No significant difference was found in the incidence of PTSD between the two arms of the study.

As we proceed to the discussion about the moral implication of MMTs, it is important to emphasize that propranolol actually does not erase memories; it counteracts the enhancing effect of anxiety associated with trauma on memory retention. Therefore, persons taking propranolol after a traumatic event will not actually forget the event, but their memory will be degraded to basically their normal level of remembering. Other agents, such as benzodiazepines, can cause amnesia affecting recall of a particular period of time and cannot be used for the erasure of specific memories.[3]

5.3 Remembering Yourself

There are a number of different ways in which commentators have approached the issue of the ethical permissibility of the use of MMTs. Leon Kass (2003) has condemned the use of medicine for enhancement in general and has also condemned the use of pharmaceutical intervention for the erasure of memories. Kass (2003) argues against memory erasure because he believes there is an obligation to have an appropriate moral response to terrible events. "Witnessing a murder should be remembered as horrible; doing a beastly deed should trouble the soul (Kass 2003)." He also condemns the pursuit of "psychic tranquility" through pharmaceutical means as immoral. It is not clear whether Kass explicitly means to condemn the use of propranolol in the prevention of PTSD or whether he means his disapprobation to apply only to cases where erasure is not treatment for psychiatric disorders. Kass illustrates his point by using the example of witnessing a murder, which is precisely the kind of event that could cause PTSD. Of course not everybody who experiences a traumatic event will develop PTSD. Thus, it is not possible to know if the use of propranolol would be therapeutic or nontherapeutic shortly after the traumatic event, the time when propranolol might be beneficial. In effect, Kass's moral condemnation of memory disrupters might obligate survivors of traumatic events to refrain from using propranolol even if it could prevent PTSD.

Other commentators have condemned the use of MMTs as well. Sandel (2004) has argued against striving for perfection as we become more able to control human traits. Striving for perfect memory would be one way in which we should not attempt to control our own capacities. Sandel's condemnation of human attempts to master physical and emotional reactions is based on the religious supposition of "the giftedness of life." Manninen (2006) does not argue directly against the use of memory enhancers, but does argue against what she calls "the gratuitous use of psychoactive medication." The drugs she discusses are mostly for the treatment of

---

[3] A study by Jin-Hee Han et al. (2009) achieved the circumscribed erasure of a single memory using a toxin to destroy particular neurons. The study was done on mice.

depression, but the conclusion could be extended to the use of memory disrupters in cases where the individual does not have a psychiatric or neurological disorder.

Liao and Sandberg (2008) have been more congenial in their normative assessment of the use of memory enhancers and disrupters. They go through a careful evaluation of potential moral issues with the use of memory enhancement, including how the use of those might affect personal identity. Liao and Sandberg argue that it is permissible to use MMTs because personal identity is resilient and does not rely on particular memories. I plan to focus precisely on this issue.

Liao and Sandberg (2008) address the problem of truthfulness as it relates to personal identity. Insofar as memories provide us with the record of past events, altering memories might change what we believe to be true about our past. Changing memories might alter how we judge our own past actions as well as the actions of the protagonists of our memories. The way we remember things might also be formative of our narrative identity. Both good and bad experiences contribute to the formation of our values and preferences and might in fact form our personality traits. Thus, memory insertion or erasure could destabilize the basis that helps form our identity. Liao and Sandberg (2008) astutely point out that the truthfulness of our memories is questionable, as was evident from the research by Loftus and Sheen presented in the previous section of this chapter. They further argue that narrative identity is more resilient and does not rely on the accuracy of a limited number of memories.

In a piece on the same topic, Liao and Wasserman (2007) argue that truthfulness of memory, however, ought to be an important consideration when it comes to the permissibility of the use of MMTs. It might be important for a soldier to remember the way he felt when he killed an enemy in combat. It might be crucial for him to remember that he lusted after the killing, but this aspect of his memory might be affected by the use of memory disrupters after the fact. Such tampering with memory could reduce our self-knowledge, especially if there are not others who could either corroborate or dispute our memories of the event. It might be difficult of course to ever have witnesses to individual experiences or emotions, particularly as those experiences are not always manifested behaviorally. Perhaps that might be even further reason to value our personal memories of events; we are the only ones to witness them from our perspective.

Given the prominence of truthfulness in the above accounts, I will address this issue. There are two kinds of truthfulness converging together in the moral evaluation of MMTs. One is about the role of truthfulness as it applies to how an event is remembered, for example, whether the person driving a car that caused the car crash had curly or straight hair. But there is also the issue of remembering our thoughts and reactions as the event was occurring. Liao and Sandberg (2008) as well as Liao and Wasserman (2007) point out that memory can provide the evidentiary basis for our interpretation of the events as well as the basis for self-knowledge. Both of these, they argue, are important and can count against the permissibility of the use of MMTs.

Let us discuss the issue of the knowledge of events first. Consider the numerous studies presented in the previous section, which showed that memories could be affected by the presentation of false information. Although such experiments involved deliberate attempts to mislead by presenting false information, those deliberate attempts reveal the vulnerability of memory in everyday life. This process of contamination probably happens with relative frequency, especially if two or more people who witnessed the event discuss the details of what was observed. Assume further that the two people rehashing the event are different in their ability to remember events; one person has good memory while the other does not. In this case, the presence of other individuals to corroborate what was seen could have a detrimental effect, and diminish the accuracy of the memory for the person who is endowed with good memory.

Consider, for example, the question that influenced the success of recall in Loftus et al's. (1989) experiment: "Did the blue car that drove past the accident have a ski rack on the roof?" Such a question is highly plausible if two people are having a conversation about a witnessed event. The false memory of the blue car can then influence the memory of the event of the interlocutor who perhaps correctly remembered the car as being green. In this case, instead of improving the veracity of memory, exchanging experiences could have a detrimental effect on the accuracy of memories. Loftus herself points out that discussing witnessed events with other people is an opportunity for memory contamination. She in fact advises that people write down their memory of the event before attempting to discuss it with anybody else (Loftus 2003). Alternatively, the false memory of the event of one person might contaminate the remembrance of the entire group. A process like this might occur in families where each family member recalls things differently and the family lore is shaped and contaminated by these different interpretations. An example of such a process is evidenced by the Sheen et al. (2001) study, where siblings were unable to distinguish accurately between what happened to them and what happened to their twin.

As was presented earlier, there are doubts about whether Loftus's interpretation of retroactive interference is accurate. McCloskey and Zarazoga challenged her substitution hypothesis. Even if we take the alternative account of biased-guessing proposed by McCloskey and Zaragoza, according to which the memory is not actually erased by the misinformation, rather, it creates a response competition that results in a recall error; the conception of memory as a reliable record of the past is nonetheless undermined. We can then wonder to what degree our memory is ever a truthful record of the past even without the use of MMTs.

There is also the more fundamental issue of the degree to which our perception of events, which then results in the memory of the event, is accurate. Expectations play a role in how we perceive objects. A study by Puri and Wojciulik (2008) tested the effects of expectation on the facility of perception. They found that correctly cuing a person for the perception of a particular object or face made recognition of the object easier. For example, cuing a participant with the name 'Goldie Hawn' improved the speed with which they were able to recognize the subsequent photo of Goldie Hawn. The recognition of Goldie Hawn was slower if the cue was the name of a different celebrity, say Tom Hanks. The recognition was even worse if the cue was from a different category, such as a place or object. But it is not just that our perception is influenced by expectation—we perceive things against a background of beliefs about the world, which include prejudices and stereotypes. Bargh and Chartrand (1999) argue that social perception is largely automated and that it can be greatly influenced by stereotypes. Thus, the perception of a person's behavior is in part based on the kinds of traits we are likely to attribute automatically to him or her. If stereotypes and expectations can affect the accuracy of perception, such inaccuracies will become part of our memories.

Let us consider now the claim that memory is a reliable source of self-knowledge. Here I intend to move from discussing memory of the event itself to the memories of our experience of those events. I plan to tackle the aspect of memory Liao and Wasserman (2007) mentioned in their example of the soldier who lusted after killing. Presumably our ability to remember how we felt about particular events can be a source of self-knowledge, which in turn could provide us with a basis for our identity. What I will say here is predictable, given that I have already spent some time discussing the failings of memory for particular events. There is reason to think that memory of past emotions is fallible. In a review of the literature on emotional memory, Christianson and Safer (1996) concluded that "there are apparently no published studies in which a group of subjects has accurately recalled the intensity and or frequency of their previously recorded emotions." This conclusion can be explained in part by appealing to the workings of the "psychological-immune system," so termed by Gilbert et al. (1998). A person recovers emotionally from a traumatic event by making sense of it (Gilbert et al. 1998,p. 637). The negative event becomes normalized and its emotional impact is less likely to be remembered accurately. Past good or bad events become less extraordinary because of the ameliorating effect of the psychological immune system.

In another review article about autobiographical memory, Wilson and Ross (2003) argue that our current self-concept influences how we remember our personal past as well as how we interpret our past selves. We tend to prefer our current self, and are more likely to denigrate our prior stages of self. Moreover, we tend to exhibit a self-serving bias that leads us to conclude that our current self is better than all our past selves. In sum, given the inaccuracy of our memory of past events as well as the unreliability of emotional memory, how we remember things does not provide a very good evidentiary basis for our sense of self. Thus, changing our memories will not deprive us of our identity as Kass had intimated. It seems indeed that if there is any such thing as a consistent sense of self, it is unlikely to be based on our autobiographical memory.

I will now turn to the issues of continuity of self over time and how it might relate to memory. In the previous chapter on cognitive enhancement, I have put forward the argument that a lot of faulty judgments about the use of cognitive enhancers stem from the failure to distinguish between numerical identity and narrative identity. Criteria of identity that rest on the establishment of numerical identity seek to determine whether a particular person is one and the same across time without relying on individual concepts of self. Narrative identity on the other hand designates exactly the aspect of personal identity that rests on conceptions of self and relies entirely on each person's sense of continuity. Because numerical identity aims to establish criteria of identity across time that do not take into account concepts of self and narrative identity rests on a person's concept of self, in those two accounts, loss of identity or discontinuity of self is established differently as well. Although Liao and Sandberg (2008) accurately distinguish between the two senses of personal identity and mostly focus on narrative identity, their assessment is that particular memories could indeed make a difference to our concept of self. I disagree with that assessment and argue that it rests on a covert endorsement of a memory criterion of numerical identity.

There are distinct ways of formulating a memory criterion. Memory criteria that potentiate the maintenance of continuity entail that an individual remember herself in the past and feel a sense of continuity between her past and present self. This type of criterion seems to most closely mirror Locke's brief description of maintenance of self through time, as described in Chap. 5 (Locke 1995 ed.). It seems that this kind of sense of identity could be maintained in spite of the loss of particular memories. One can maintain a sense of continuity even if one forgets aspects of one's personal history. This way of prioritizing memory for maintenance of identity is more compatible with the notion of narrative identity because it relies on the person's individual judgment of continuity of identity. An alternative way of formulating the memory criterion is that maintenance of identity over time requires the maintenance of particular memories. For example, the person who has Marie Antoinette's memories, caused in the right way through experience, is Marie Antoinette (Perry 1978). And if Marie Antoinette forgets important aspects of her life, she will cease to be herself. This is an attempt to establish numerical identity over time by designating memory and the maintenance of particular memories as essential for survival.

It seems to me that this second type of memory criterion is the only sense in which the use of MMTs could affect identity. If it is true that we must have a certain set of core memories in order to maintain a sense of self, then loss of those memories could change who we are. There are some very conspicuous flaws with this view, one of which is accounting for identity over time for those individuals who have memory disorders. Based on this criterion of identity over time, it would be difficult to account for the fact that amnesiac Marie Antoinette is still one and the same person. But even in less dramatic cases, this kind of identity criterion is not adequate. Given the inaccuracy of autobiographical memory and the degradation and reinterpretation of our memories over time, conditioning identity on the maintenance of memories seems unwarranted because normal changes in memory could produce discontinuity in identity.

Perhaps this problem could be avoided by designating a certain cluster of memories as core memories whose maintenance over time would be required for the survival of a person. Certainly, a memory criterion cannot require that a person have eidetic memory in order to maintain identity over time. In such a case, the use of MMTs would usurp identity if they are used to erase core memories. Traumatic memories, given their presumed pervasive influence on a person's life, might be just the kind of memories that would qualify as core memories. If that were the case, then perhaps one could support the idea that the use of MMTs merits serious reconsideration because it might change identity.

If one were able to devise a memory criterion for personal identity and select the types of memories that constitute the core memories required for personal continuity, then a person could be deemed to have changed identity even if she feels that her narrative identity has remained continuous. Based on the concept of numerical identity, one could have normative criteria by which one could say that a person without certain core memories is no longer one and the same. For example, the Marie Antoinette who no longer remembers saying "Let them eat cake!" is no longer Marie Antoinette. Using the concept of narrative identity, however, we would not have any independent way of fixing Marie Antoinette's identity, save for her own impression of her continuity of self. Thus, when it comes to narrative identity, we rely on the individual to tell us which memories are most relevant to her sense of self. Although others remember Marie Antoinette's famous remark and think of it as revealing of her identity, she might not rank that memory as one of her core memories. Therefore, loss of that memory for her would not cause discontinuity of identity.

Narrative continuity of self is based on the person's own feeling that such continuity exists. Given that narrative identity is subjective in such a way, the importance of particular memories is entirely based on the person's own ranking of those memories. If an individual feels that certain memories are fundamental to her sense of self, she might not decide to undergo any memory modification that could put such important memories in jeopardy. Now if a person wishes to erase memories because they are not constitutive of her sense of self, there is no recourse to argue that she should not because it will rob her of her identity. It might be that traumatic events, albeit prominent in memory, are not fundamental to a person's sense of self, and use of MMTs could facilitate the process of forgetting memories that might be nevertheless unduly exerting their influence on one's sense of self. Interpreted in this manner, MMTs could help promote a sense of narrative identity, not jeopardize it.

Another way of assuaging the fear that persons using MMTs will forget who they are is by pointing to further reasons to think that a sense of self can exist independent of the persistence of autobiographical memories. Klein and Nichols (2012) present a compelling example of a patient, R.B., who as a result of being involved in a motorcycle accident had experienced serious head trauma. R.B. had retrograde and anterograde amnesia for events that occurred soon before and after his accident. R.B. had an intact episodic memory for earlier events from his past. Episodic memory is long-term memory of personal experiences that are specific to time and place, for example, the memory of one's first kiss. Curiously, R.B. reported remembering those events as if they had happened to somebody else; the memories lacked a sense of "miness." Nonetheless, R.B. reported a continuity of his sense of self. Although interpretation of this case is complicated, one could argue that it shows that a sense of self is not dependent on the maintenance of autobiographical memories at all.

5.4 Authenticity

Personal identity and authenticity are sometimes used interchangeably, but they can also be conceived of as related but distinct concepts. While personal identity designates a certain set of properties as important for a concept of self, authenticity requires living in accordance with one's true self, based on a concept of self. Thus, it is worth having a separate discussion about the impact of MMTs on authenticity. In this section, I try to address the criticism mounted by Alexandre Erler (2011) against DeGrazia'a view, which he characterizes as an existentialist view of authenticity. DeGrazia's view on authenticity is formulated as follows: "A autonomously performs intentional action X iff (1) A does X because she prefers to do X, (2) A has this preference because she identifies with and prefers to have it, and (3) this identification has not resulted primarily from influences that A would, on careful reflection, consider alienating" (Erler 238). Erler characterizes the account he prefers as a *true self* account. Based on Erler's account, authenticity is the virtue of being faithful to one's true self, when doing so is intrinsically valuable (Erler 2011, p. 238). The true self account is superior, according to Erler, to all the other accounts of authenticity because it can justify the charge of inauthenticity in a greater number of cases, especially in cases where inauthenticity could be the only risk of using MMTs.

In the true self account, memory editing can interfere with authenticity by threatening truthfulness. Persons may forget who they are and what they did, which would interfere with their being able to be truthful about their identity. Furthermore, MMTs might interfere with a person's actual dispositions to act in a certain way, which also could lead to inauthenticity. Erler argues that inauthenticity should be one of the factors considered when evaluating the use of MMTs because authenticity has intrinsic value.

Erler assures us that he does not wish to fall into the trap of confusing numerical identity with narrative identity and agrees with DeGrazia (2005) about the distinction between those two concepts. Erler argues that the true self is a central feature of narrative identity; it includes personality traits, character traits, personal preferences, self-image, moral and religious preferences. But he disputes DeGrazia's view that most of those traits are part of our idea of true self only if we identify with them; some aspects of our narrative self are true, whether or not we identify with them. Given that identification with one's core traits is not a requirement, then truthfulness about self would necessitate that there be an independent criterion for establishing core personality traits. In my view, this aspect of Erler's position is a holdover from the conception of identity as numerical because such criteria also aim to identify persons with their enduring psychological traits.

To illustrate his argument, Erler describes the case of Oscar. Oscar is a 20-year-old gay man who is unhappy with his sexual orientation. Oscar attempts to extinguish his desire for the same sex and finds a therapist to help him in that quest. After many years of unsuccessfully striving to become heterosexual, Oscar's therapist suggests that he enroll in a clinical trial for a new chemical treatment of homosexuality. Oscar enrolls and is "cured" of his homosexuality. Because Oscar does not endorse his homosexuality during his entire life, then, Erler argues, DeGrazia would have to argue that being homosexual was never part of Oscar's true self. But based on Erler's view, Oscar's homosexuality is a central feature of his narrative identity, regardless of whether he identifies with it. By denying his homosexuality, Oscar is failing to be truthful about his sexual orientation and is living an inauthentic life. Although one's sexual orientation might indeed be an immutable characteristic, there is no reason to assume that sexual orientation must be an aspect that each person chooses to include in a core sense of self. We all have a number of personality traits, character traits, and preferences, but not all of them are part of our true self. In other words, just because something is true of Oscar, it does not mean it has to be true of his core self.

Erler argues that his version of authenticity requires that person's true self include central features of narrative identity that significantly shape and form him or her, including traits based on ways in which people see and treat him or her. Thus, Oscar should include his sexuality as part of his central characteristics because of the ways in which other people think of him and because of the ways other people treat him. But surely this cannot be right. One could presume that one of the reasons why Oscar works so hard to change his sexuality is precisely because of the way in which other people see him and the ways in which other people treat him. One could further presume that given the lingering prejudice against homosexuals, Oscar wishes to be treated better and because of this he attempts to change an aspect of himself that predisposes him to be discriminated against. If Oscar lived in a more congenial environment, where his sexuality was not part of the way in which other people see him, he would probably not try as hard to change that aspect of his person. Moreover, assuming a world without prejudice against those who are homosexual, Oscar would be less likely to think of his sexuality as one of his more notable aspects.

There are a number of other cases where there is a disparity between how other people think of an individual and that individual's self-conception. Because of the influence social context exerts on the ways in which we conceive of ourselves, society might in fact erroneously influence our true self and prevent us from being authentic. Often others prioritize what we might consider unimportant features of our person, for example, race, nationality, gender, and even, in some cases, religious affiliation. Emphasizing the importance of one's gender or nationality, for example, could be a sign of respect, but it could also be a form of discrimination if the way in which a person is thought of and treated is solely or mostly in terms of race, gender, nationality, and so forth. Moreover, individuals who are discriminated against because of those features might be forced to focus on traits of their person in ways that might actually distract from their authenticity. Therefore, insistence on a criterion other than personal endorsement might cause inauthenticity.

Erler, I presume, wishes to capture the intuition that there must be some facts about narrative identity that are independent of the ways in which we might think of ourselves. And perhaps there are some things that are true of each person independent of what they think of themselves, but those could be used in a criterion of numerical identity, as I argued in Chap. 5. A criterion of numerical identity, if successful, applies to persons in the same way whether or not they countenance those features as parts of their core self. Furthermore, the existence of such facts is not an argument that they must be included into the person's conception of a true self. In other words, although there might be facts about human psychology or even facts about the personality traits of a particular person, those facts are not facts about a person's true self. Narrative identity is an attempt to tell a story about who we are, and it requires the first-person ranking of personality traits. If narrative identity is conceived as a first-person way of forming identity, then the person who ought to be the judge of true self is the individual.[4]

There is a further difficulty with Erler's requirement of truthfulness, which is the indeterminacy of one's true self over time. Oscar's case rests on the selection of a personality trait that does persist over time. But there are many more that do not. Parfit's example (1984) of the young Russian nobleman who is liberal in his youth but becomes conservative with age is a realistic example, given the results of the study by Quoidbach, Gilbert and Wilson (2013). The Russian nobleman in his youth was liberal and disillusioned with his privileged status in society and wished to give money to the peasants. In his old age, as he had anticipated, he changed his mind; he changed what he believed about the nature of serfdom and adopted different values, which made him more conservative. We can assume that this change in values and beliefs was not imperceptible and that others noticed the change in the Russian nobleman, which caused them to think of him differently and to treat him differently. Both in DeGrazia's sense and in Erler's sense, the Russian nobleman has changed his true self.

Erler argues that his objection to authenticity does not rest on the distinction between artificial and natural ways of enhancing well-being by changings one's self (Erler, 246). He would question attempts to change one's core self even if they came through meditation or any other, so-characterized, natural way of changing oneself. The young Russian, however, changes spontaneously without effort. The change in his personality was neither untruthful nor was it the result of active attempts to change his core self. It seems then that spontaneous changes of self do not violate authenticity even by Erler's standard.

The change of self of the Russian nobleman illustrates a very noticeable turnabout in values and preference, but more subtle changes occur to most people over time. These changes also can occur spontaneously without medication or meditation. During the various stages of self where people incrementally change who they are, there are times when the person's self will be ambiguous. Imagine Mary, who, because of a difficult childhood, has a tendency to act aggressively in certain situations. Mary might over time spontaneously become less prone to temperamental outbursts and accept difficulty in life in a more temperate manner. Because the

---

[4] This is not to argue that we are responsible only for those features of ourselves that we endorse. One is responsible for being a murderer, liar, or cheater even when we do not regard those traits as part of our selves.

change in her personality is gradual, during it, she might in some circumstances react like her old self and in other instances like her new, less temperamental self. Given this equivocation between the old and the new self, we could say that her true self is undetermined. Mary is not the only one who undergoes such changes of self: over time most of us change who we are, and our lives are marked by periods where our true self cannot be fixed. Moreover, if different aspects of the self change at different times, personal characteristics are most likely always in flux. If I am correct in characterizing the self as not determinate, then one cannot argue that there are fixed facts about one's true self.

Based on DeGrazia's account, where the first-person endorsement of one's true self is required, one could maintain that a person has a determinate true self insofar as she thinks that there are core characteristics that are currently true of her narrative identity. Therefore, actual fluctuations in personal characteristics can be countenanced because the person herself might always have a sense of continuity of her true self. But, given Erler's rejection of the personal-endorsement criterion of a true self, one cannot be sure which characteristics can be properly said to be true of me at each stage of my life, and my true self cannot be determined.

Returning to the use of MMTs and their impact on authenticity, Erler argues not only that there are ways of fixing certain aspects of our true self that are independent of the person's own identification with those traits, but that maintaining authenticity is intrinsically virtuous. Thus, in some cases, either enhancing or editing one's memories is wrong because it might cause inauthenticity. Erler, however, admits that the commitment to authenticity is only prima facie, and that there might be cases where the use of MMTs would be justified.

A case in which the use of MMTs is not justified is where compromising authenticity is not outweighed by the benefits of forgetting some painful memories. Consider the example of Elizabeth, who was socially ostracized in high school. Elizabeth was not popular in high school; the other girls in the class made efforts to exclude her from social interaction. Elizabeth has since become a successful and happy person, with a good career, and happy family life. Nonetheless, she still holds a grudge. She keeps her distance from her high school classmates and refuses to go to reunions. Although Elizabeth's grudge is not a major impediment to her quality of life, she decides to use MMTs to blunt the emotional impact of the social exclusion she felt in high school. In this case, Erler argues that the use of MMTs might not be justified. He argues that by using MMTs, Elizabeth is suppressing her actual tendency to hold a grudge when mistreated (Erler 2011, p. 245).

Erler argues that if Elizabeth decides to erase her memory of the events from high school, she would be changing an aspect of herself in a way that would make her inauthentic. Elizabeth has a tendency to hold grudges and is not a person who reacts in an easygoing manner. In considering this case, let us briefly discuss the plausibility of such erasure. First, the erasure of a single memory would not be enough to rid Elizabeth of her unpleasant memories from her youth. She probably has a number of memories accumulated during her time in high school. Daniel Dennett (1981, p. 44), in an evaluation of the possibility of memory insertion, argues that maintenance of biographical and logical coherence would require that an insertion of one false belief, say, that I have a sister living in Tulsa, be accompanied by the insertion of a multitude of other beliefs. The insertion of the false memory that I have a sister would have to be accompanied by the induction of beliefs about my sister's name, her age, profession, memories of growing up together, and so forth. Similarly, the excision of one particular memory would require the removal of a web of other memories and beliefs in order for memory erasure to be therapeutic for Elizabeth.

Moreover, the erasure of Elizabeth's episodic memory for certain events might not be enough to extinguish the grudge she feels towards her classmates. People learn facts from particular events that can then be abstracted from that context and applied generally to other situations. Thus, erasing particular episodic memories would not extinguish the memories of declarative facts and nondeclarative facts learned from those events. If childhood events are erased, Elizabeth could be left with a free-floating, but persisting, grudge toward her high school peers that would be hard to interpret. Moreover, if holding grudges is part of Elizabeth's repertoire of dispositions or personality traits, erasure of memories for particular events that elicited such a response from Elizabeth would not eliminate her general tendency to react in just that way. Hence, since erasure of particular memories will not jeopardize behavioral dispositions, it therefore will not rob persons of their authenticity.

Erler also suggests the possibility of Elizabeth's deciding to blunt her memories by using propranolol. Propranolol would not be an option for Elizabeth for a number of reasons. Beta-blockers are efficacious only if used immediately after a traumatic event, and Elizabeth has been out of high school for many years. The events of Elizabeth's youth, as described by Erler, might not qualify as traumatic enough. If Elizabeth was bullied, then her experiences might qualify as traumatic. Assuming that Elizabeth's experiences were traumatic enough and that she had the choice of taking propranolol to blunt the impact of those events, then her memory of those events would be diminished, but not erased. What remains in her memory might suffice to exert some kind of emotional impact that could support the maintenance of Elizabeth's grudge. If Elizabeth's high school experiences were not traumatic enough, then propranolol might not help, as it has not been studied for its effects on unpleasant memories that are not traumatic. We know that propranolol does not have an effect on memories for neutral events (Reist et al. 2001).

Ignoring the actual limitations of current MMTs, let us consider Elizabeth's decision to erase her memory and the impact on her authenticity. As I argued before, persons might have a number of different characteristics and those might be true of them whether or not they wish to include them into their core self. Although Elizabeth has a tendency to hold grudges, that might not be the way in which she thinks of her true self. Imagine that Elizabeth believes that she was easygoing for most of her youth until she encountered the mean girls in her high school. Those events then made her more cynical and suspicious of other people. Elizabeth now thinks that propranolol might help her return to her old easygoing self by blunting the impact of her unwanted memories. MMTs would then promote Elizabeth's authenticity and, using Erler's view, the promotion of her authenticity would be intrinsically virtuous.

In Chap. 4, I mentioned the study by Bolt and Schermer (2009), where patients treated for Attention Deficit and Hyperactivity Disorder (ADHD) reported a variety of perspectives when it came to the influence of medication on their sense of personal identity. Although some study participants expressed that taking medication felt like it changed their personality, some reported that they felt more like themselves (Bolt and Schermer 2009). One could also imagine that those individuals who suffer from PTSD do not think of their traumatic memories as being constitutive to their sense of self, but feel them to be an impediment to feeling like their true self. In that case, the use of propranolol or any other efficacious memory-modifying technology could be the way to promote one's true sense of identity, and the use of them would be justified not just because of improvements in well-being, but because of the promotion of authenticity.

5.5 Conclusion

In this chapter, I consider the ethical implications of memory-modifying technologies, especially as they pertain to personal identity and authenticity. Normative condemnation of MMTs often relies on the assertion that memory is veridical and can be used to ground personal identity. I have presented a range of studies that substantiate the claim that autobiographical memory is unreliable. Given this unreliability, I have argued that memory cannot be used as an evidentiary basis for either knowledge of past events or self-knowledge. Therefore, MMTs that alter autobiographical memory will not jeopardize personal identity. This especially true if one remains consistently committed to the notion of narrative identity. Due to its flexibility, narrative identity can be maintained despite changes in memory. This type of flexibility in an account of identity is particularly fortuitous because of the number of changes in personality and memory that occur over a lifetime.

In addition, I have shown that authenticity could only be fixed based on personal endorsement of one's core characteristics constituting one's true self. Because of the spontaneous fluctuations of each person's character traits, values, and preferences, I argue that one's core true self cannot be determined in a manner that does not rely on a person's own estimation that certain traits are part of her true self. Moreover, I dispute the argument that MMTs can challenge an individual's authenticity. In fact, the use of pharmaceutical agents, such as propranolol, might be justified because they promote authenticity in cases where traumatic or unpleasant memories are obstructing the person's ability to live an authentic life.

References

American Psychiatric Association. (2000). *Diagnostic and statistical manual of mental disorders* (4th ed., text rev.). doi:10.1176/appi.books.9780890423349

Bargh, J. A., & Chartrand, T. L. (1999). The unbearable automaticity of being. *American Psychologist, 54*(7), 462–479.

Bolt, I., & Schermer, M. (2009). Psychopharmaceutical enhancers: Enhancing identity? *Neuroethics, 2*, 103–111.

Cahill, L., & McGaugh, J. L. (1996). Modulation of memory storage. *Current Opinions in Neurobiology, 6*, 237–242.

Cahill, L., Prins, B., Weber, M., & McGaugh, J. L. (1994). Beta-adrenergic activation and memory for emotional events. *Nature, 371*, 702–704.

Caine, E. D., Weingartner, H., Ludlow, C. L., Cudahy, E. A., & Wehry, S. (1981). Qualitative analysis of scopolamine-induced amnesia. *Psychopharmacology (Berlin), 74*(1), 74–80.

Christianson, S. A., & Safer, M. A. (1996). Emotional events and emotion in autobiographical memory. In D. Rubin (Ed.), *Remembering our past: Studies in autobiographical memory* (pp. 218–243). Cambridge: Cambridge University Press.

DeGrazia, D. (2005). Enhancement technologies and human identity. *Journal of Medicine and Philosophy, 30*, 261–283.

Dennet, D. (1981). *Brainstorms: Philosophical essays on mind and psychology.* Cambridge, MA: The MIT Press.

Elliot, R., Sahakian, B. J., Matthews, K., Bannerjea, A., Rimmer, J., & Robbins, T. W. (1997). Effects of methylphenidate on spatial working memory and planning in healthy young adults. *Psychopharmacology, 131*, 196–206.

Erler, A. (2011). Does memory modification threaten our authenticity? *Neuroethics, 4*, 235–249.

Gilbert, D. T., Pinel, E. C., Wilson, T. D., Blumberg, S. J., & Wheatley, T. P. (1998). Immune neglect: A source of durability bias in affective forecasting. *Journal of Personality and Social Psychology, 75*(3), 617–638.

Han, J. H., Kushner, S. A., Yiu, A. P., Hsiang, H. L., Buch, T., Waisman, A., et al. (2009). Selective erasure of a fear memory. *Science, 323*(5920), 1492–1496.

Henry, M., Fishman, J. R., & Younger, S. J. (2007). Propranolol and the prevention of post-traumatic stress disorder: Is it wrong to erase the "sting" of bad memories? *The American Journal of Bioethics, 7*(9), 12–20.

Hoge, E. A., Worthington, J. J., Nagurney, J. T., Chang, Y., Kay, E. B., Feterowski, C. M., et al. (2012). Effect of acute posttrauma propranolol on PTSD outcome and physiological responses during script-driven imagery. *CNS Neuroscience & Therapeutics, 18*, 21–27.

Izquierdo, I., Bevilaqua, L. R., Rossato, J. I., Lima, R. H., Medina, J. H., & Cammarota, M. (2008). Age-dependent and age-independent human memory persistence is enhanced by delayed post-training methylphenidate administration. *Proceedings of the National Academy of Sciences, 105*, 19504–19507.

Kass, L. R. (2003). *Beyond therapy: Biotechnology and the pursuit of human improvement.* Washington, DC: The President's Council on Bioethics.

King, D. J. (1992). Benzodiazepines, amnesia and sedation: Theoretical and clinical issues and controversies. *Human Psychopharmacology: Clinical and Experimental, 7*, 79–87.

Klein, S., & Nichols, S. (2012). Memory and the sense of personal identity. *Mind, 121*(483), 677–702.

Liao, S. M., & Sandberg, A. (2008). The normativity of memory modification. *Neuroethics, 1*, 85–99.

Liao, S. M., & Wasserman, D. T. (2007). Neuroethical concerns about moderating traumatic memories. *American Journal of Bioethics, 7*(9), 38–40.

Locke, J. (1690, 1995 ed). *An essay concerning human understanding* (pp. 174–190). London: Everyman Press.

Loftus, E. (1977). Shifting human color memory. *Memory and Cognition, 5*, 696–699.

Loftus, E. (2003). Our changeable memories: Legal and practical implications. *Nature Reviews. Neuroscience, 4*, 231–234.

Loftus, E. F., & Palmer, J. C. (1974). Reconstruction of automobile destruction: An example of interaction between memory and language. *Journal of Verbal Learning and Verbal Behavior, 13*, 585–589.

Loftus, E., Donders, K., Hoffman, H. G., & Schooler, J. W. (1989). Creating new memories that are quickly accessed and confidently held. *Memory and Cognition, 17*, 607–616.

Manninen, B. A. (2006). Medicating the mind: A Kantian analysis of overprescribing psychoactive drugs. *Journal of Medical Ethics, 32*, 100–105.

McCloskey, M., & Zarazoga, M. (1985). Misleading postevent information and memory for events: Arguments and evidence against memory impairment hypothesis. *Journal of Experimental Psychology: General, 114*, 381–387.

Mehta, M. A., Owen, A. M., Sahakian, B. J., Mavaddat, N., Pickard, J. D., & Robbins, T. W. (2000). Methylphenidate enhances working memory by modulating discrete frontal and parietal lobe regions in the human brain. *Journal of Neuroscience, 20*(6), RC65.

Muller, G. E., & Pilzecker, A. (1900). Experimentalle beitrage zur lehre vom gedachtnis. *Zeitschrift fur Psychologie, 1*, 1–300, as cited in Green, R. L. (1992). *Human memory: Paradigms and paradoxes*. Hove/London: Lawrence Erlbaum Associates, Publishers.

Parfit, D. (1984). *Reasons and persons*. Oxford: Clarendon.

Perry, J. (1978). *A dialogue on personal identity and immortality*. Indianapolis: Hackett Publishing Company.

Pitman, R. K. (1989). Post-traumatic stress disorder, hormones, and memory. *Biological Psychiatry, 26*, 221–223.

Pitman, R. K., Sanders, K. M., Zusman, R. M., Healy, A. R., Cheema, F., Lasko, N. B., et al. (2002). Pilot study of secondary prevention of postraumatic stress disorder with propranolol. *Biological Psychiatry, 51*, 189–192.

Puri, A., & Wojciulik, E. (2008). Expectations both helps and hinders object perception. *Vision Research, 48*, 589–597.

Quoidbach, J., Gilbert, D. T., & Wilson, T. D. (2013). The end of history illusion. *Science, 339*, 96–98.

Reist, C., Duffy, J. G., Fujimoto, K., & Cahill, L. (2001). Beta-adrenergic blockade and emotional memory in PTSD. *International Journal of Neuropsychopharmacology, 4*, 377–383.

Sandel, M. (2004). The case against perfection. *The Atlantic Monthly, 293*(3), 51–62.

Sheen, M., Kemp, S., & Rubin, D. C. (2001). Twins dispute memory ownership: A new false memory phenomenon. *Memory and Cognition, 29*, 779–788.

Sheen, M., Kemp, S., & Rubin, D. C. (2006). Disputes over memory ownership: What memories are disputed? *Genes. Brain and Behavior, 5*(Suppl. 1), 9–13.

Vaiva, G., Ducrocq, F., Jezequel, K., Averland, B., Lestavel, P., Brunet, A., et al. (2003). Immediate treatment with propranolol decreases posttraumatic stress disorder two months after trauma. *Biological Psychiatry, 54*, 947–949.

Wade, K. A., Garry, M., Read, J. D., & Lindsay, S. A. (2002). A picture is worth a thousand lies. *Psychonomic Bulletin and Review, 9*, 597–603.

Wasserman, D. (2004). Making memory lose its sting. *Philosophy & Public Policy Quarterly, 24*, 12–18.

Wilson, A., & Ross, M. (2003). The identity function of autobiographical memory: Time is on our side. *Memory, 11*(2), 137–149.

Chapter 6
Brain Imaging and the Privacy of Inner States

Abstract Improvements in our ability to identify brain function as it is occurring through brain imaging have brought to the forefront the issue of mental privacy. Several authors have cited potential infringement on privacy as one of the primary ethical issues related to the application of brain imaging technology to clinical, research, and legal contexts. I challenge the argument that the use of functional magnetic resonance imaging (fMRI) poses a threat to mental privacy and that this type of privacy requires extra protections. I review all the positions about the nature of mental states that establish a category of mental privacy and conclude that none of those views can support both the claim that there is a category of mental privacy and that this type of privacy can be violated through the use of brain imaging. I further argue that the only position about the nature of mental states that erases the epistemological gap between introspection and third-person access to our inner states is eliminative materialism. Eliminativism, however, does this by denying the categories of folk psychology, including the category of mental privacy. Finally, I argue that because no view about the nature of mental states can support the argument that 'brain reading' will result in 'mindreading,' fMRI does not pose a threat to mental privacy. I conclude that special protections for mental privacy are not required because informational privacy already protects, at least in principle, the privacy of information about patients and about research participants in whatever way it is obtained.

6.1 Introduction

The advent and development of brain imaging technology, specifically functional Magnetic Resonance Imaging (fMRI), has advanced the localization of psychological states and processes in the brain. The increased ability to associate discrete pockets of brain function with distinct psychological processes has promoted the idea that mental, or psychological, states can be explained entirely in terms of brain processes. Brain imaging has been used for different purposes and in several disciplines. Functional MRIs are used in psychology to discover how a normal brain produces psychological function. They have been used in studies to identify the differences between normal brain function and the brain functioning of those afflicted with a neurological disorder, such as Parkinson's (Weiller et al. 2006). In

© Springer Science+Business Media B.V. Dordrecht 2016   95
N. Gligorov, *Neuroethics and the Scientific Revision of Common Sense*,
Studies in Brain and Mind 11, DOI 10.1007/978-94-024-0965-9_6

the domain of neuroscience of ethics,[1] fMRI are used to identify the instantiation of moral behavior in the brain (Greene et al. 2001). Further, fMRI have been used for the localization of single thoughts as they are realized in the brain. In a recent study by Kay et al. (2008), it was shown that a computer could correctly identify the object a person was thinking about, for example, a house or a car, based entirely on an fMRI of that person's brain activity. Also, fMRI is increasingly being considered as a method of lie detection in criminal proceedings.[2]

The improvements in our ability to identify brain function as it is occurring and the use of this technology to help localize particular thoughts have prompted discussions about mental privacy. Several authors have raised infringement on privacy as one of the primary ethical issues related to the application of brain imaging technology to clinical, research, and legal contexts. Illes and Racine (2005) have compared thought privacy to genetic privacy, and argue that thought privacy might merit even more protection than genetic privacy. Meegan (2008) has argued that the use of brain imaging technology can give rise to an Orwellian-type dystopia where the Thought Police could have access to others' mental states. Richmond (2012) argues that the privacy of our mental states is inherent in our folk psychology, and because of that, the potential intrusion into that private realm is morally repugnant to some.

In this chapter, the emphasis is on whether fMRI will affect privacy in the clinical and research setting. Primarily, I focus on the assessment made explicit by Richmond—and implicit in the calls for the special protection for mental privacy— that, based on our folk psychology, mental states are inherently private. An argument is needed to establish that information about our psychological states is private in a different way from all the other information routinely obtained during the course of research and treatment. I countenance that brain imaging can be used to obtain private information about an individual, but I argue against the claim that the information obtained would be different from information regularly obtained from patients and research participants. This is because I argue that fMRI cannot be used to reveal the subjective aspects of our inner states, i.e., *what it is like* for an individual to have a thought or sensation. To do this, I draw on established approaches in philosophy of mind on the nature of mental states and on the relationship between mental and physical states. I utilize those accounts to show that none of them can be used to support the claim that information about the subjective aspects of mental states can be obtained using brain imaging technology. In other words, there is no such thing as mindreading technology, which I do not take to be a temporary limitation of brain imaging. I see this conceptual point as being of practical importance because if it is true that intrusion into the mental realm is morally unacceptable to some, categorizing brain imaging as mind reading, and referring to it in this manner, might discourage research participation and even prevent the incorporation of brain imaging into clinical care.

---

[1] A branch of neuroethics identified in Roskies (2002).

[2] For a thorough review of this possibility, see Meegan (2008).

The argument in this chapter will be presented in four parts. In Sect. 6.2, I present the current practical and conceptual limitations of the fMRI technique. This part of the chapter presents what is only a temporary challenge to the claim that brain imaging is a threat to mental privacy. In Sect. 6.3, I present my argument that fMRI cannot pose a threat to mental privacy in principle. I consider four ways in which mental states can be said to be inherently private. I argue that all those ways depend on views that maintain that there are subjective facts principally unobtainable through brain imaging. In Sects. 6.4 and 6.5, I present views that do not characterize mental states as having subjective aspects and I show that only eliminative materialism (EM) could be used to argue that one could obtain the same type of information about inner states from introspection as from brain imaging, but that is because EM denies folk-psychological mental categories. In effect, the endorsement of EM implies the rejection of the category of mental privacy. Hence, no view about the nature of mental states can be used to both establish a category of mental privacy and to support the argument that brain imaging poses a threat to that type of privacy. If the ethical issue is then recast as the potential for brain imaging to infringe on brain privacy, I argue, in Sect. 6.6, we already have a concept of privacy that can be used to protect the privacy of information obtained through brain imaging for research and medical treatment. Thus, I maintain that no special protections for thought or mental privacy are required because there is no argument to support the view that information about brain processes is any more private than information obtained about any other body part.

6.2   The BOLD Signal fMRI[3]

In the following section, I aim to describe some basic facts about functional MRI as well as to highlight some of the limitations of this technique for capturing brain activity. The aim here is not in any way to discredit the fMRI technique nor to question its general use in psychology, neuroscience, or any other field. Rather, it is an attempt to familiarize the reader both with some of the complexities involved in obtaining fMRI results and with some of the complexities in interpreting them. In the process, I wish to make salient the number of inferential steps necessary to support the claim that fMRI captures brain activity,[4] which should in turn temper the strong claim that fMR images are a direct measure of brain activity.

Magnetic resonance imaging (MRI) is primarily used to produce structural images of organs, including the central nervous system. MRI provides only a picture of the affected area, but does not provide direct information about its functional ability. An MRI is similar in principle to a mammogram: it is a way of visualizing what is inside the body, without having to open it physically (Crawford 2008).

---

[3] This section of the chapter is based on a previously published manuscript by Gligorov and Krieger (2010).

[4] For more on this, see Roskies (2008).

Functional neuroimaging, such as a functional MRI (fMRI), on the other hand, is primarily an attempt to capture information about brain function.

Yet the graphics that both MRI and fMRI produce require immense complexity of the mechanical, physical, and analytic procedures utilized in creating them (Logothetis 2008). FMRI is not a direct measure of neural activity, but a technology that functionally maps the working brain by tracking changes in oxygenation in particular brain regions. This, in turn, is accomplished through measuring regional blood supply in the brain, and correlating these regions with various behavioral functions and cognitive tasks (Fenton et al. 2009). When neurons are active, they consume oxygen, carried by hemoglobin in red blood cells from local blood vessels and capillaries. The local response to this increased oxygen demand is an increase in blood flow to regions of increased neural activity, occurring after a delay of several seconds. Although active neurons consume oxygen and thus increase the amount of deoxygenated hemoglobin in the blood, the increased supply of oxygenated blood results in a net increase in the concentration of oxygenated hemoglobin (Roskies 2008).

Oxygenated and deoxygenated blood yield different magnetic signals, which can be detected using the MRI scanner. The magnetic resonance (MR) signal of blood therefore varies depending on its level of oxygenation. The blood-oxygen-level dependent (BOLD) signal is the fMRI measure of blood deoxyhemoglobin, and thus the BOLD MR signal uses the blood oxygenation level as a surrogate marker for increased neural activity. Increased levels of deoxyhemoglobin reduce the BOLD signal; reduced concentrations increase it. Almost all fMRI research uses BOLD as the method for determining where activity occurs in the brain. BOLD effects are measured using rapid acquisition of images, which can capture moderately good spatial and temporal resolution; images are usually taken every 1–4 s, and the voxels (three-dimensional pixels, or volumetric pixels) in the resulting image typically represent cubes of tissue about 2–4 mm on each side in humans. Once this data is acquired, it is statistically analyzed to generate an "image" that is used to visualize the location of discrete brain areas from which activity is recorded.

Functional neuroimaging studies result in enormous data sets that must first be parsed into what is and is not valuable. This requires setting thresholds on the raw data, a process that is inherently somewhat arbitrary, as well as hypothesis-driven. It is not just threshold-limited data from one scan that is typically analyzed in order to make claims about neural activity, but it is pooled and processed data from multiple trials, and often, multiple subjects. Although this inserts a further layer of abstraction, this pooling is necessary because the signal-to-noise ratio for neuroimaging is quite low; data from multiple scans is averaged in an attempt to maximize the signal being studied (Roskies 2008). There are numerous debates regarding the proper statistical and data analysis techniques that should be used in fMRI studies, ranging from questions about how to correct for multiple comparisons to whether analysis should be hypothesis-driven or whether brute-force statistics suffice (Roskies 2008).

The fMRI data are not originally in the format of an image, but in data structures that encode numerical values of MRI signal intensity collected in an abstract framework called *k*-space. Visual representations of data in *k*-space bear no visual resemblance to images of brains (Roskies 2008). These data are transformed into a spatial format through a mathematical formula called a Fourier transform, resulting in an image that can be color coded and presented atop a typical grayscale MRI image of the corresponding brain. The design decisions are made by convention, but a number of analytic decisions are employed in the creation of fMRI images, including setting thresholds, smoothing out the voxels, and choosing colors to indicate particular findings of the study in question. Some experimenters use color gradations to indicate relative levels of activity, while others use color gradations to indicate relative levels of statistical significance.

There are some notable limitations to the fMRI technique. The limitations of fMRI pertain to both *how* it identifies brain function, and *what* it identifies as brain function. In terms of *how* fMRI identifies regions of brain activity, critical factors determining the utility of fMRI for drawing conclusions in brain research are spatial and temporal resolution (Logothetis 2008). The spatial resolution of fMR imaging is not as refined as that of traditional structural MRI. While neuroimaging allows visualization at the millimeter scale, the incremental building blocks of regional neural activity are those that occur on a cellular and subcellular level. Although these may ultimately be the most elemental and important phenomena that generate brain function as we know it, fMRI is not able to resolve events occurring on this microscopic scale. Since functional neuroimaging is intended to identify regions responsible for the generation of behaviors, attempting to study dynamic interactions at the level of single neurons would probably make little sense, even if it were technically feasible, particularly considering there are 10^{10} neurons in the cortex alone (Logothetis 2008). However, given that the size of an fMRI voxel is on the order of several cubic millimeters, each voxel comprises approximately five million neurons. The degree of spatial resolution an imaging modality must have in order to be useful necessarily depends on the question being addressed: "It makes no sense to read a newspaper with a microscope," as neuroanatomist Valentino Braitenberg pointed out (as quoted in Logothetis 2008). Systems for recording individual nerve firings might miss the "big picture," and neuroimaging that captures the whole brain perhaps neglects relevant small-scale neuronal activity.

Similarly, neural function is necessarily a real-time process, and the temporal resolution at which fMRI captures brain activity is limiting as well. Traditional fMRI experimental paradigms have excellent functional contrast-to-noise ratio (they can identify the signal recorded during a behavioral test from the "quiet" during between-test periods), but they are usually long intervals, lasting from 20 to 60 s, and may be confounded by the general state of arousal of the subject. High-speed fMRI methods, capable of whole-brain imaging with a temporal resolution of a few seconds, enabled the employment of more modern experimental designs. The time course of the response in such experiments is closer to the underlying neural activity (Logothetis 2008). Brain scans are nonetheless not images of mental activity in process, as the neural activity on a cellular level occurs on a time scale orders of magnitude faster than the BOLD response can measure. Neuroscience here is limited, not only by the practical limitations of the fMRI technique but on the incomplete understanding of how the fleeting, oscillatory electrophysiology at the cellular level gives rise to the large-scale patterns of brain activity that we can recognize. The limitations of fMRI as an abstracted, large-scale surrogate for actual microscopic neural activity render it similar to any other physiological correlate of mental function, such as the electrodermal response used in a conventional lie detector (Uttal 2001).

A further limitation regarding *how* fMRI is utilized to identify local areas of brain activity is the use of subtraction analysis. Experimental fMRI paradigms require a task state designed to place specific experimental demands on the brain, as compared with a control state. A number of theoretical assumptions are required for this methodology; in particular, that by subtracting the brain activity recorded in the control state from that recorded during the task state, the difference between the two states can be identified and correlated with the specific cognitive demands of the task. This paradigm requires that the task and control states differ in a single cognitive process, which is often difficult to prove. Furthermore, it presumes a somewhat linear form of brain processing, such that serial subtractions would identify, rather than obfuscate, the neural mechanisms of the cognitive processes under investigation. Functional MR imaging may fail to distinguish other physiologically relevant parameters, such as large changes in the firing rate of a few neurons, small changes in the firing rates of many neurons, or changes in temporal pattern of nerve cell activity in the absence of changes in mean firing rates (Roskies 2008). Distinguishing background noise from important, yet subtle, signals is as yet an imperfect science. Without a coherent unifying theory of brain function, it is difficult to know what data are being subtracted that should not be, and what data are not being subtracted that should.

An additional limitation of the fMRI technique pertains to subtraction analysis and the conceptual failure to take into account aspects of brain function that are not discretely localizable. In an attempt to make a controlled experimental environment to identify a single brain function, the subtraction method eliminates from the picture the fact that, as in real life, the entire brain is active in both conditions. This can yield an artificial impression of neat functional localization, which subtracts out all the distributed functions (Crawford 2008). Although more modern imaging protocols have attempted to correct this issue, overlapping networks of neurons subserving different functions are likely to go unnoticed, owing to the spatial averaging that characterizes the subtraction paradigms (Logothetis 2008).

A final limitation that will be mentioned is with regard to what fMRI identifies as active areas of the brain. As described above, fMRI uses oxygenated blood flow as a surrogate marker for cell nerve functions. Neuronal firing can be both excitatory and inhibitory, and the vast networks of neural impulses that ultimately yield brain function rely on a complex interplay between excitation and inhibition. Indeed, shifting the balance from one of excitation to inhibition (or the reverse) is the mechanism behind myriad neurological conditions, from seizures to migraines to mood disorders. Dynamic alterations in this balance, whether they lead to net excitation or inhibition, inevitably and strongly affect the regional metabolic energy demands, and thus the regulation of oxygenated cerebral blood flow (Logothetis 2008). Although the functional implications of excitatory and inhibitory synaptic transmission are quite different, BOLD fMRI fails to distinguish between them (Roskies 2008), and indicates only that *something* is happening in a region of increased oxygenated blood flow in the brain.

One of the most widely discussed future uses of functional neuroimaging technology (including fMRI) is for lie detection. Indeed, there are several commercial fMRI enterprises that offer high-tech lie detection services based on research comparing neuronal activation patterns of liars and truth-tellers (Fenton et al. 2009). The technique of fMRI lie detection is based on the functional imaging of the cognitive process required for deception. The basis of fMRI lie detection is the finding that more fMRI activation is seen in the prefrontal and anterior cingulate regions in the "lie condition" relative to the "truth condition" in an experimental setting (Fenton et al. 2009).

Although the specifics may vary considerably with the circumstances, to lie always requires the intent to deceive. Deception involves knowing (or at least believing something to be) the truth, and saying or implying the opposite, and it necessarily involves judgments about the beliefs and knowledge of the audience (Pardo and Patterson 2010). Brain imaging lie detection is predicated on the idea that lying requires more cognitive effort, and therefore more oxygenated blood, than truth telling. Particular areas of the brain are said to be deciding when and whether to lie, and then engaging in the processes to carry out this decision (Pardo and Patterson 2010). The message put forth in the popular media is that brain scanning promises to show us directly what the polygraph showed us obliquely (Talbot 2007). But as with traditional polygraphs, neuroscience research is looking for a correlation between deceptive behavior and some other, objectively measurable phenomena. With polygraphs, it was increased heart rates, breathing, and perspiring; with functional neuroimaging, it is increased blood flow to certain regions of the brain.

A central worry is whether the behaviors of research participants in test situations properly qualify as lying (Fenton et al. 2009). The limitations on applications of neuroimaging technology are further exacerbated by the need for cooperation on the part of the individual being imaged, and the need to rely on first-person reporting. In Fenton et al.'s view, the anticipated breakthroughs in "brain reading" (i.e., imaging and interpreting brain activity) in the service of "mindreading" (i.e., ascribing specific beliefs, desires, thoughts, and intentions) imply an independence from the cooperation of the imaged individual that is unrealistic (Fenton et al. 2009). Functional neuroimaging indeed cannot be accomplished without the extraordinary cooperation and patience on the part of the subject. Such willing participants in fMRI research projects, having consented to participate in research, knowing that their thoughts are being analyzed with a brain scan, would appear to have opposite goals (utilizing different mental functions, presumably with different neurological substrates) than people in real-world situations attempting to conceal the truth.

All the limitations mentioned above should be used to qualify any claims that fMRI can be reliably used for "brain reading," as well as the claim that brain imaging can currently be used to circumvent introspection and first-person reports of psychological states. Of course, the current limitations of the fMRI technology are not really what matters when evaluating the argument that brain imaging could pose a threat to mental privacy in the future. We could imagine that brain imaging techniques as well as our knowledge of brain processing in general will progress, and we should evaluate even the potential of fMRI to pose a threat to mental privacy in the future. For that reason, in the next few sections I plan to present a philosophical argument that evaluates the possibility that brain imaging could infringe on mental privacy. I will set aside the caveats presented in this section and assume that brain imaging can capture brain activity as it is happening. The argument I will make, however, should prevent the inference that "brain reading" is enough for "mindreading."

6.3   Different Kinds of Mental Privacy

Mental states in everyday parlance are often characterized as private. The sense in which our psychological states are private can be interpreted in several different ways depending on the type of view endorsed about the relationship between mental states and physical states. Descartes argued that we know our minds first and our bodies second, and that this primacy of access confirms the intimate link between the self and the mind. Descartes argued that mental states are tied to our conscious access to them, which means that only the person experiencing them can access them. Moreover, conscious access to mental states reveals their properties to the person experiencing them in a veridical way. Based on this view, we cannot be wrong about features of mental states since their nature is determined by how they seem to us. Descartes's view of mental states produces a contrast between the mental and the physical because mental states are defined as having characteristics opposite to those of physical states. This contrast between the mental and the physical in turn prevents mental states from being incorporated into a physical theory of brain processes. It would seem clear that any such view of mental states would preclude the claim that brain imaging technology could provide access to mental states because it denies that mental states can be reduced to patterns of brain activity.

Nagel (1975), like Descartes, defines mental states in terms of our conscious access to them, rendering aspects of mental states subjective. Nagel argues that even if we had exhaustive knowledge of facts about a creature's perceptual mechanism, say, that of a bat, and all the facts about its physiology, we would still fail to know *what it is like* for that bat to have perceptual experiences. There are a number of philosophical terms used to designate the experiential aspect of perception and other psychological states, i.e., 'what it is like for one,' some of those terms include phenomenal character, qualitative state, or qualia. Although distinction can be drawn between those terms, I will use them interchangeably in this chapter to designate the subjective aspects of mental states. Jackson (1986) also argues that a complete physical theory would fail to cover facts about mental states because facts about our experience contain nonphysical facts. On Jackson's and Nagel's view, the epistemological issue of how mental states are known or accessed give rise to a metaphysical issue about the nature of mental states. David Chalmers (2003) characterizes these as epistemic arguments and formulates their general form as follows:

1. There is an epistemic gap between physical and phenomenal truths.
2. If there is an epistemic gap between physical and phenomenal truths, then there is an ontological gap, and materialism is false.
    Thus,
3. Materialism is false.

A broad category under which we can join Descartes's view and those of Nagel and Jackson is that of dualism, either of substance or of property. These views solidify the mental as essentially private because the nature of the mental is characterized in terms of the subjective access to those states. Descartes's argument that we know our mind better than our body and Nagel's argument that mental states are inherently subjective can support the view that mental states contain private information about an individual. If the way in which an individual experiences the world—what it is like for that individual to be that individual—is strictly linked to a subjective perspective, then facts about what it is like for an individual to be in a particular mental state are inherently private. This is very different from the privacy of our body, which depends on purposeful concealment of bodily parts or on the contingent fact that many organs are not readily observable. The development of requisite technology can help reveal the aspects of the body that are inner in this sense. But if one adopts the view that subjective facts about mental states are accessible only introspectively, no technological advances could replace introspection and be used to acquire those facts. In what follows, I will show how characterizing mental states as subjective enough to warrant a category of mental privacy, undermines the claim that 'brain reading' will result in 'mind reading.' In other words, one cannot maintain both that mental states are inherently private and that brain imaging could be used to infringe on that type of privacy.

To begin, I will avail myself of A. J. Ayer's (1963) characterization of different types of mental privacy.[5] A. J. Ayer (1963) distinguished four ways in which mental states have been said to be private: incommunicability, requiring special access, unsharability, and incorrigibility. I will begin with incommunicability. Incommunicability derives from the assumption that mental states are ineffable, and cannot be adequately described by language. Based on this account, our descriptive reports of mental states cannot accurately describe the character of mental states, and aspects of mental states not captured by language remain private.

---

[5] I should note that the presentation of Ayer's view in this chapter is neither an endorsement of the notion of mental privacy as described, nor a claim that Ayer endorsed the Cartesian notion of privacy. In fact he develops the argument only to criticize it. I use him merely to illustrate the view of mental privacy that is entailed by a nonreductive view of mental states.

Even without being a dualist, one could maintain that mental states are ineffable or nonconceptual, in the sense that they might not be reportable through the use of language. For example, one could argue like Tye (1995) that bodily sensations such as pain have nonconceptual content. This means that sensations might represent the world or the body as being a certain way: for example, pain represents a particular body part as being in a disordered state. But this kind of representation need not have a linguistic structure required for the possession of the concept of pain. Babies and animals have pain, but do not have the ability to report those feelings using language.

Alternatively, Block (2004) argues that bodily sensations are sometimes nonrepresentational. For example, the experience of a red tomato represents the tomato as red. But the qualitative aspects of the red sensation—what it is like to see red—do not represent anything; they do not attribute any properties to the world. Qualia, according to Block, do not have intentional content; they are not about tomatoes or anything else in the world. If aspects of our bodily sensations do not have intentional content, they cannot be captured by language and are ineffable. If they are ineffable, then they are basically not communicable by means of language.

Going back to Kay et al.'s (2008) study, where functional brain imaging was used to identify a particular pattern of brain activity associated with a particular thought, say, a thought of a house or a car, the fMRI was used to individuate the thought, based on its intentional content. The thought was about a house or a car, and based on brain activity alone, Kay et al. were able to know that a person was thinking of a house or a car. But the fMRI would not be able to bridge the gap between the representational content of the thought, i.e., that the thought was about a particular object, and the qualitative aspects of the experience. The use of the functional MR imaging can enable a researcher to have access to the representational content of my experiences, but not to what it is like for me to have them. Insofar as the qualitative aspects of experience are what mark them as mental, because they create the gap between physical and phenomenal facts, fMRI cannot be used to circumvent introspection since it cannot bridge that gap.

In his attack on functionalism, Block (1980) argued that two individuals who are functional duplicates would be exactly alike except for their qualia—what it is like for those individuals to be in a particular mental state. However, he does countenance that the difference in the qualitative aspects of their inner states could be captured by EEG or any other technology that could access brain function as it is happening. Functional imaging could be that kind of technology. Thus, although looking at an fMRI of a person seeing red would not be able to convey what it is like for that person to feel a sharp pain, it can be used to know that a person is undergoing a particular kind of qualitative experience. Aspects of mental states would still remain private because only the person having them would have knowledge of its qualitative features, for example, what it is like to feel a sharp pain.

A second sense of mental privacy as formulated by Ayer, refers to the notion of special access. A mental state is private if the person having that state is able to know about it in a way different from that of other people. This sense of privacy relies on the idea that only the person having a mental state has direct access to it and everybody else can only inferentially know that state by observing behavior or listening to first-person reports. An fMRI can be said to circumvent the need for inference. For example, in the Kay et al. (2008) study, if a person is having brain activity associated with a thought that p, we can know that a person is having a thought that p without requiring a verbal report.[6] Descartes (2003 ed.) argued that it must be possible for the mental substance and the physical substance to interact with each other, and he even suggested that this is done through the pineal gland. Extrapolating from that very limited claim, one could argue that Descartes could countenance that mental activity could in some way cause brain processes recorded by an fMRI. But this is not enough to support the conclusion that having a brain process is like having a mental state, and that accessing one is accessing the other.

Based on a view that is not physicalist, at least some aspects of mental states would not be realized in the brain in order to justify positing either an additional mental substance or mental properties that remain unaccounted for by a physicalist account describing the nature of brain states. Even those who are not dualist, but maintain that aspects of mental states are ineffable or that qualitative aspects of mental states overflow their representational content, would have to argue that intro-spective access is special in that only the person having mental states can have direct access to the qualitative aspects of those states. Although functional MR imaging could circumvent the need for verbal reports about the aspects of mental states that are representational and reportable, some aspects would remain private even if we assume that every property of our mental states is in some way realized by the brain.

A third sense of mental privacy, as formulated by Ayer, is unsharability. In this sense, mental states are private if no one else can have them as I do. This sense combines the idea both of incommunicability and of special access. If I am not able to communicate my occurrent mental states and I am the only one who can access them in a way that reveals their properties, then my mental states are not sharable with others. Rorty in "Incorrigibility as the Mark of The Mental" (1970) argues that this type of mental privacy can be challenged by the conceivability of telepathy or science fiction scenarios in which people can communicate their mental states directly using some future brain scope. But even if it became possible, by some scientific advancement, to share mental states directly, it is not obvious that the act of sharing a mental state from one person to the other would not alter that mental state to the point that the original experience would remain private. What it might be like for Jane to experience George's experience of red might be different from what it might be like for George to have that experience.

Imagine, for instance, that George is a fashion designer who is knowledgeable about colors. He knows many more names for colors than the average person and has been exposed to a greater number of hues than average Jane. Imagine that Jane and George are friends, and imagine further that technology has been developed that could allow people to share mental states directly, rendering the sharing of photo-graphs and videos obsolete. George in Paris sees a red vase that he admires, and

---

[6] In Chap. 7, I describe imaging studies that could enable researchers or clinicians to know that person is in pain based on patterns of brain activity, obviating the need for a verbal report of pain.

wants to share his experience of the red color of the vase with his friend Jane in New York. He decides to avail himself of this new technology, which can operate even over distances; there are experience-sharing booths just as there used to be telephone booths.

Using the experience sharing booth, George sends Jane his experience of the red vase. Jane receives the experience of red, but, unlike George, has a very limited range of color experiences. Once she receives George's experience of the color red, it becomes incorporated into her color palette. Jane's color palette is different from George's, and the color of the vase is experienced by her differently. If the qualitative aspects of the particular shade of red are set in comparison with other hues of red, say, based on the similarities and differences between the different hues, the qualitative aspects of each hue will be slightly different from person to person depending on their exposure to different shades of red. George and Jane would have different phenomenal concepts of red. Thus, even a direct sharing of experiences of red will not result in the same qualitative experience, if two people are dissimilar in relevant ways.

The fourth and final sense of mental privacy, as formulated by Ayer, is the incorrigibility of mental states. Rorty (1970) argues that mental states are incorrigible because certain knowledge claims about them cannot be overridden. In other words, reports of mental events cannot be shown to be false. For example, if a person is reporting that she is in pain, and if there are not reasons to believe that the person is lying, it seems there is no recourse to challenge her report of pain. According to Rorty, incorrigibility is the mark of the mental and can be used to distinguish mental events from physical processes. Kripke (1980) provides an additional argument for the incorrigibility of mental states; he maintains that there is not an appearance and reality distinction when it comes to sensations. If it appears to Jane that she is in pain, it is the same as Jane being in pain. If for an individual to be in pain, it is enough that it seem to her that she is in pain, then even if fMRI can be used to identify the locus of pain states in the brain,[7] subjective reports of pain would still be incorrigible.[8] Assuming a subjective characterization of pain, in a case of conflicting reports, where the fMRI indicates that an individual is not in pain while the person says that she is in pain, the verbal report would supersede the imaging report.

Based on nonreductive views of mental states, for all the reasons covered in this section, mental states are deemed private in all the four ways cited by Ayer (1963). This would be true even for those who think that mental states are physical but not identical to a particular type of physical state. If mental states are not entirely reducible to physical states, then the qualitative aspects of mental states cannot be captured by fMRI or any other conceivable brain scope. Consequently, mindreading will not be among the sequelae of brain imaging technology.

The upshot of this section is not that brain imaging cannot be used to obtain private information, only that it cannot be used for mindreading. A criticism of my

---

[7] For an example of such a study, see Coghill, R.C., McHaffie, J. G., and Yen, Y.-F. (2003).

[8] Chapter 7 of this book is devoted to the subjectivity of pain. In that chapter, I argue that the incorrigibility of mental states is false.

approach in this section is that it does not address the primary concern about the use of fMRI, which is that it can be used to obtain information that a person is thinking and it can be used to discover what the person is thinking about, as per Kay's study (2008). I will address this issue in Sect. 6.6. For now, all I wish to accomplish is to establish that there is not mindreading technology.

6.4 Mental Privacy and Physicalism

There are a number of broadly construed physicalist explanations of mental states. Such positions include Eliminative Materialism (EM), type-type physicalism, token-token physicalism, and functionalism. Of those four, I will show that only EM supports the claim that fMRI can provide a third-person alternative to introspection.

According to type-type physicalism, a type of mental state, say, pain, is identical to a type of brain state. This view follows the rules of strict identity; therefore any property a mental state has is also a property that a brain state has and vice versa. Type physicalism is not based on the denial of mental features; rather, it is based on a successful reduction of mental to physical states. Thus, a science about the brain should be able to capture all the relevant features of mental states. In order to claim that an fMRI could provide an alternative to introspection, one would have to not only have a successful reduction of mental to physical states, but to have a successful reduction of how we speak of mental states to how we speak of brain processes, or to use Place's terminology, we would have to achieve a successful reduction of 'mental talk' to 'brain talk' (Place 1956).[9] In case of such complete reduction, Quine and Rorty, among others, have argued that mental talk becomes superfluous. Quine (1969) says that to define something is in effect to eliminate it because the entities that really exist are the ones to which we have reduced mental states. In effect, this argument is that if reduction is achieved, mental categories, including mental privacy, are illusory. This last type of reductivism is a precursor to contemporary versions of eliminative materialism, which will be discussed later in this section.

Token-token physicalists argue that a particular occurrence, or token, of a mental state, say my current pain in the leg, can be reduced to a particular token of a brain state. But it is not true, in this view, that a type of mental state, for example any pain state, can be reduced to a particular type of physical state. It could be true that each time I am in a pain state, I am also in a physical state, but it is not true that that physical state will always be the same even for a particular individual. Although this might seem as a congenial solution, there are some problems with it, particularly for the claim that localization of brain states using brain imaging could provide an alternative to first-person introspection. In a seminal article, Jerry Fodor (1974) argues that if we admit that there are homogonous psychological kinds, with

---

[9] Place (1956) argued against the possibility and the need for the reduction of 'mental talk' to 'brain talk.'

heterogonous physical realizers, for example, if the state of pain can be realized differently in humans, aliens, robots, or silicon brains, then although we could have laws that establish generalities between psychological states, we cannot establish laws between different levels of explanation, such as pains and their physical realizers. Although there might be psychological laws and neuroscientific laws, there are not bridge laws connecting the two types of explanation. In other words, although it might be true that there could be local reductions of human pains to brain states, or alien pains to alien brains, or silicon pains to silicon brains, the disjunction of human pains, alien pains, and silicon pains and the disjunction of their physical realizers—human brains, alien brains or silicon brains—does not constitute a natural law. Fodor takes this to be an argument against intertheoretic reduction and supportive of the autonomy of the special sciences (Fodor 1974, 1997). Fodor's argument makes it possible to claim that generalities established by the special sciences simply cannot be reduced, and perhaps do not even exist, at a lower level of explanation; which is in effect an argument that mental categories, such as beliefs, thoughts, and sensations do not exist at the level of neuroscientific explanation. This type of nonreductive functionalism precludes the inference that brain imaging can result in mindreading.

There are reductive forms of functionalism according to which functional analyses of psychological states are thought of as sketches of fully developed mechanistic explanations. Once the structural information omitted from the functional analyses are filled out, those become mechanistic explanations of neural systems (Piccinini and Carver 2011). But even such a view does not support the claim that fMRI could supplant introspection. Functionalism is an account of the nature of mental states, which, even if reductive, could countenance that there is an epistemological gap between knowing mental states introspectively and knowing facts about those states from a third-person account. A reductive functionalist can countenance an epistemological gap between qualitative aspects of sensations and their representational aspects, without admitting that those qualitative facts are nonphysical facts. Hence, none of the physicalist and even reductivist views presented in this section thus far can support the conclusion that fMRI can be used to obtain all facts about mental states available to introspection.

Eliminative Materialism (EM), as espoused by Paul Churchland (1992), however, could be used to support a view that fMRI can provide an alternative access to mental states. According to EM, properties of mental states that cannot be successfully reduced are not real and should be omitted from any theory covering human psychology. Churchland's argument is that the subjective phenomenal properties of a color sensation are objective properties of the brain, which can be de facto identified with a type of brain activity in the visual cortex (Churchland 2005). Churchland argues that one can reconceptualize, using a neuroscientific theory, in order to individuate brain states as inner states. Experiencing a red sensation is to introspect "various spiking frequencies in the nth layer of the occipital cortex" (Churchland 1992, p. 65). Churchland argues that eliminating sensations and reconceptualizing to a neuroscientific framework would bridge the gap between physical and phenomenal facts. An neuroscientist with complete propositional knowledge of color perception who has not seen the color red, has not had her nth layer activated, but she would be able to predict what it would be like to be in that brain state based on what she already knows about color.[10] Because this person would be able to predict what it is like to see red, she would not learn any new facts after she sees her first red tomato, and the gap between physical facts and phenomenal facts would disappear. Based on this view, if there is any sense in which sensations are private, it is not because they are subjective, but because they are experiences happening inside the body, like all of our other bodily states.

Furthermore, eliminativism of this sort is a broad attack on all the categories of folk psychology, including propositional attitudes such as thoughts and beliefs. Churchland's argument against these categories is that the sentence-like structure of thoughts and beliefs is not representative of how the brain actually functions. Research into the neural structures that underlie the organization and processing of perceptual information reveals that such processes accomplish a great variety of complex tasks, some of which show complexity far in excess of natural language (Churchland 1992, p. 19). Churchland like Fodor seems to be arguing that thoughts and beliefs do not exist in the brain, although Fodor takes this to indicate the independence of psychology from neuroscience, while Churchland concludes that those categories do not exist because they are not supported by neuroscience.

Assuming this view, brain imaging technology could be used to identify all facts about a person's inner states. This result, however, was achieved by denying the all our mental categories, including the subjectivity of mental states and their inherent privacy. EM can be used to support the claim that brain imaging can be used to gain information about brain function, but not for 'mindreading.' Thus, no established views about the nature of mental states can maintain both the claim that mental states are inherently private and the claim that all facts about mental states can be captured using brain imaging.

6.5 Not Mental, But Neural Privacy

In the previous sections, I established that no view about the nature of mental states could support the argument that brain reading will result in mindreading, but eliminativism can be used to support the view that brain imaging technology could, at least in principle, be used to obtain the same type of information as introspective access to mental states. I will now assess, using Ayer's model, whether any of the arguments for the inherent privacy of mental states could be applied to brain or neural privacy, preserving the argument that the privacy of our brain states is inherent and requires special protections. I will conclude that it cannot.

---

[10] My argument here is based on Churchland's view about the normal individual's ability to predict what it would be like to experience colors outside the normal visual experiences including chimerical qualia. For more see Churchland (2005).

As a reminder, Ayer mentions four ways in which mental states are purportedly private: they are not communicable; they are accessed in a special way; they are not sharable; and are incorrigible. It was argued that mental states are incommunicable because natural language cannot capture all facts about them. Based on the eliminativist view of mental states, one ought to be able to capture all aspects of our inner states with propositional statements plus know-how. Propositional statements capture facts about the world by utilizing sentences in a language, while know-how entails knowledge acquired though experience, such as learning how to ride a bicycle. One could argue that know-how could be incommunicable. Knowledge of skills is not straightforwardly communicable through propositions; one needs to ride a bicycle to acquire that ability. This inability to share know-how, however, does not derive from the inherent subjectivity of the experience of riding a bicycle. Any true fact about our inner states can be couched as an objective property in the world (Churchland 1992).

A similar argument can be used to counter that claim that mental states are private because they are accessed in a special way. This special access, i.e., introspection, is only available to the person having that state. As I have shown in the previous section, Churchland has argued that physical states can be introspected. He has done so by denying that there are any such things as subjective phenomenal properties; rather, he argues that subjective mental states are nothing but objective phenomenal states of physical things. If one considers redness as on objective property of a cherry, a person looking at a cherry is merely observing and then reporting an objective property in the world rather than reporting on a subjectively experienced qualitative state. Furthermore, if phenomenal properties are objective and physical, when we report introspecting mental states, we are in fact introspecting brain states. If that is what phenomenal properties are, then any privacy based on the subjective aspect of first-person experience dissipates.

I have, also, discussed the feature of unsharability of mental states. Given that this feature appears to be a combination of incommunicability and special access to mental states, it will suffice to say that an eliminativist would not be able to accept unsharability of mental states, for reasons cited above. Moreover, based on the eliminativist view, there would be no principled objection to science fiction scenarios by which mental states could be shared directly from person to person. The possibility of such an exchange would be an empirical claim, which could turn out to be either true or false depending on further discoveries in neuroscience, scientific psychology, and the development of requisite technology.

Finally, to an eliminativist, incorrigibility would not be a feature of mental states because first-person reports would no longer be the final arbiter of the occurrence and features of mental states. Scientific discoveries in neuroscience could contribute to our conception of mental states and in many respects correct our views about human psychology and properties of mental states. As mentioned in Sect. 2.2 of Chap. 2 of this book, advancements in scientific psychology have shown that introspective access is not always the most accurate method of accessing mental states. If we are inclined to incorporate scientific knowledge into our theories of mental states, as an eliminative materialist would, then incorrigibility is not a feature of our inner states. In fact, an elimantivist would argue that neuroscience provides the most accurate characterization of brain states, and would not give credence to the claim that introspection provides incorrigible access to our inner states.

To recapitulate my current line of argument, in order to claim that brain imaging can be used as an alternative to introspection, one would have to adopt the eliminativists recasting of that concept, which is that introspection does not reveal any subjective facts about our inner states, only objective facts about our brains. By adopting such a line of argument, however, one surrenders the ability to claim any of the special features of mental privacy because those features exist only based on views that characterize mental states as having subjective features. And without the subjectivity of mental states, brain privacy seems more like a contingent fact about brain states. Brain states are inner states. By 'inner,' I mean to designate the fact that mental states are inside the skull, and have been difficult to record and observe directly before the advent of the Electroencephalogram (EEG) and then of functional MRI. Furthermore, the problems often associated with the precision and accuracy of fMRI might be resolved with the further development of such technology, thus removing the practical barrier that has kept brain states private and only accessible by the person having those states. If the privacy of neural states derives solely from the fact that brain states are inner, then they are not different from the state of any other organ that is inside the body and can be observed only through the use of biomedical technology. If neural privacy is like bodily privacy in this sense, then our increased ability to access brain states poses privacy concerns more akin to informational privacy as it applies to bodily states. Informational privacy focuses on privacy of the information about a person and leaves open the kinds of ways and places that such information can be obtained.[11] In sum, the notion that mindreading is amongst the ethical implications of brain imaging should be abandoned because no view about the nature of mental states allows us to both endorse the purportedly special features of the mental and the materialist claim that brain imaging could be used to access those features.

6.6 Informational Privacy

The thrust of my argument from the previous sections is that the type of information that can be obtained from brain imaging, fMRI in particular, is similarly private as other information that is routinely obtained from individuals either through clinical care or in the course of research. If one adopts the view that there is something inherently subjective about our psychological states and that what it is like for an individual to be that individual is private because it is subjective, then the worry that fMRI could infringe upon that privacy should have been dispelled in Sect. 6.2 of this chapter. Alternatively if one thinks that fMRI might one day become an alternative to introspective access, one is in effect denying the subjective realm of the mental

---

[11] For a representative view, see Fried (1968).

and adopting the view that mental states just are brain processes and are private in the same way as hepatic function. Either way, information obtained through the use of brain imaging can be subsumed under already existing concepts of privacy and confidentiality as they are used to protect private medical information. I suspect that folk-psychological notions of privacy will adjust to reflect the inclusion of information about brains under the notions of bodily privacy, especially once we dispel the misconception that brain imaging can be used for mindreading.

My argument that neural privacy should be considered as a species of informational privacy is not meant to undermine the importance of privacy as it is endorsed in medicine and research.[12] The duty of clinicians and researchers to safeguard privacy is protected by the law, albeit with varying degrees of success, through the Health Insurance Portability and Accountability Act (HIPAA). The conclusion that confidentiality of private information is important for both medical care and the conduct of research is well founded.[13] Moreover, ethical and legal confidentiality protections need to be adjusted to accommodate developments in medicine and biomedical research.

A possible counterargument to the position espoused in this paper is that neural privacy is exceptional because disclosure of information about brain function could be very harmful, akin to disclosure of genetic information, and this requires additional privacy protections. An argument considering the parallels between mental and genetic privacy can be found in Illes and Racine (2005). I will not evaluate the parallel between genetic and brain privacy, nor will I evaluate the strength of the claim that genetic information is exceptional because those arguments are outside of the scope of my chapter. Some examples of harm would be employment discrimination based on genetic predisposition for illness, or, in the past, health coverage discrimination because of the high risk to develop a disease that might prove costly for health insurance companies. If infringements on neural privacy turn out to be exceptionally harmful in similar ways, then neural privacy should be protected with legal and other guidelines to minimize the potential for harm.

I will now evaluate some potential harms of the use of brain imaging and show that the information obtained using this technology is similarly private, and its disclosure similarly harmful, as the information already routinely obtained in patient encounters or through research participation. For example, an fMRI could identify that a person is thinking about a particular object, e.g., a house. The Kay et al. study (2008) discussed earlier supports this possibility. One could further imagine that brain imaging could be used to diagnose psychiatric disease,[14] or that it could be used even to identify pre-symptomatic individuals who will develop a psychiatric

---

[12] For views on the nature of the right to privacy, see Rachels (1975) and Thompson (1975). For a review of the different views on privacy, see Gligorov, Nada, et al. (2013).

[13] For an argument emphasizing the importance of confidentiality, see Kipnis (2006). For an argument that confidentiality, rather than privacy, is a more apt way of describing the obligations of researchers and physicians to protect information obtained from patients and research participants, see Gligorov, Nada, et al. (2013).

[14] For a study supporting this application of brain imaging, see Bansal et al. (2012).

illness. Brain imaging could be used to identify personal characteristics; for example there are differences in areas of brain activation between those of psychopaths and normal individuals.[15] Thus, one could imagine that such markers of brain activity could be used to identify and stigmatize pre-symptomatic individuals.

But now let us consider the kinds of information that could be obtained through a regular visit to the doctor. For example, in conversation one might be asked to discuss bowel habits, sexual habits, nutritional habits, or one could be asked to discuss consumption of alcohol or other drugs. All this information is considered by many to be private and could be harmful if disclosed outside the auspices of the doctor-patient relationship. Moreover, information routinely obtained in clinical encounters could be stigmatizing to the patient. For example, patients who suffer from drug addiction might not be listed for organ transplantation or individuals who are diagnosed with a psychiatric illness might not have their decision making capacity respected (Grisso and Appelbaum 1995). One could respond that all the information in a doctor-patient visit is disclosed voluntarily, and the patient could just refuse to share information. Still, even if the patient does not share information about drug abuse or risky sexual encounters, a blood test could reveal both drug use and sexually transmitted diseases.

Perhaps all there is to the worry that brain imaging will infringe on privacy, is that such information will be obtained without consent, or despite refusal. There are two things to say about this worry. One is that it is not obvious why fMRI or other future brain scopes would be exempt from the requirement of informed consent. Most persons are able to refuse any treatment or procedure; this right is safeguarded by laws and by the adherence to the ethical principle of autonomy. The same should be the case for the use of brain imaging, the regulation of which should be subsumed under our current protections of the right to refuse treatment or to opt out of any research protocols. In fact, current uses of brain imaging require informed consent either from the patient or from a patient's surrogate.

Another issue that might impact how we evaluate the use of fMRI is that what doctors know about a patient is not limited to only that which the patient chooses to disclose. A psychiatrist does not depend on the patient's confession that she is a psychopath in order to diagnose that disorder. Patients often do not have the type of insight necessary to help in the diagnosis in such a way; in fact, people often discover new things about themselves through psychiatric diagnosis. So brain imaging could help in psychiatric diagnosis, but it would not introduce an entirely novel level of intrusion. A good psychiatrist can conclude from behavioral and other clues more than a patient can convey or even wishes to convey. Similarly, neurologists do not always need to use brain imaging, and did not for many years before the advent of that technology, to determine facts about brain function or diagnose deficits in neurological function. Thus the use of fMRI seem more like an additional way of doing what is already common practice in the medical or research setting. The information is not different in kind, i.e., it is information about brain function that

---

[15] For a study about the differences in brain function between normal individuals and psychopaths, see Birbaumer et al. (2005).

can be obtained in a a variety of different ways, some more direct than others, and the information obtained is not private in different ways. Thus, the privacy of our brains can be subsumed under the already existent categories of informational privacy.

6.7   Conclusion

The advancement of neuroscience and the development of brain imaging technology have elicited a variety of ethical concerns. One of those concerns is whether brain imaging will pose threats to mental privacy. In this chapter, I have reviewed a number of different theoretical approaches to the nature of mental states. I argue that mental privacy as defined by Ayer can be supported only by nonreductive approaches to mental states. Both substance and property dualists maintain that mental states have properties that brain states do not have; specifically they have qualitative aspects. Based on those views, those aspects cannot be captured by a physicalist account and cannot be pictured using brain scans. Thus, according to a dualist account, brain imaging cannot be seen as a threat to mental privacy. Similarly, physicalist accounts that are not reductive can also maintain that there are aspects of mental states that cannot be captured using brain imaging, thereby preserving the privacy of mental features.

The only reductive and physicalist account that could be used to support the claim that fMRI can close the gap between phenomenal facts and physical features is an eliminativist approach to the mind and body problem, which is based on the denial of mental features. Eliminativists argue that phenomenal properties of mental states are objective properties of either objects or the brain. Objectified in this manner, our sensations and thoughts are no longer private in the special sense described by Ayer. Mental states neither are incommunicable, unsharable, incorrigible, nor do they require special access. Based on this view, the privacy of our brain states is akin to the privacy of all our inner states and can be categorized as a subset of informational privacy and be accorded all of the confidentiality protections already in place to protect information about patients and about research participants.

References

Ayer, A. J. (1963). *The concept of a person*. New York: St. Martin's Press.

Bansal, R., Staib, L. H., Laine, A. F., Hao, X., Xu, D., Liu, J., et al. (2012). Anatomical brain images alone can accurately diagnose chronic neuropsychiatric illnesses. *PLoS ONE, 7*(12), e50698. doi:10.1371/journal.pone.0050698.

Birbaumer, N., Veit, R., Lotze, M., Erb, M., Hermann, C., Grodd, W., et al. (2005). Deficient fear conditioning in psychopathy: A functional magnetic resonance imaging study. *Archives of General Psychiatry, 62*, 799–805.

Block, N. (1980). Are absent qualia impossible? *The Philosophical Review, 89*(2), 257–274.

Block, N. (2004). Mental paint. In M. Hahn & B. Ramberg (Eds.), *Reflections and replies: Essays on the philosophy of Tyler Burge*. Cambridge, MA: MIT Press.

Chalmers, D. J. (2003). Consciousness and its place in nature. In S. Stich & F. Warfield (Eds.), *Blackwell guide to the philosophy of mind* (pp. 102–142). Malden: Blackwell Publishing.

Churchland, P. M. (1992). *A neurocomputational perspective*. Cambridge, MA: MIT Press.

Churchland, P. M. (2005). Chimerical colors: Some phenomenological predictions from cognitive neuroscience. *Philosophical Psychology, 18*(5), 527–560.

Coghill, R., McHaffie, J. G., & Yen, Y. F. (2003). Neural correlates of interindividual differences in the subjective experience of pain. *Proceedings of the National Academy of Sciences, 100*(14), 8538–8542.

Crawford, M. B. (2008). The limits of neuro-talk. *The New Atlantis, 19*, 65–78.

Descartes, R. (1664, 2003ed.). *Treatise of man*. Amherst: Prometheus Books.

Fenton, A., Meynell, L., & Baylis, F. (2009). Ethical challenges and interpretive difficulties with non-clinical applications of pediatric FMRI. *American Journal of Bioethics, 9*(1), 3–13.

Fodor, J. (1974). Special sciences (or the disunity of science as a working hypothesis). *Synthese, 28*, 97–115.

Fodor, J. (1997). Special sciences: Still autonomous after all these years. *Noûs*, (31) Supplement: Philosophical Perspectives, 11, Mind Causation, and World, 149–163.

Fried, C. (1968). Privacy (a moral analysis). *Yale Law Journal, 77*(1), 475–493.

Gligorov, N., & Krieger, S. C. (2010). Functional brain imaging, free will, and privacy. In S. M. Kabene (Ed.), *Healthcare and the effect of technology: Developments, challenges, and advancements* (pp. 233–251). Hershey: IGI Global Publishing.

Gligorov, N., Frank, L. E., Schwab, A. P., & Trusko, B. (2013). Privacy, confidentiality, and new ways of knowing more. In R. Rhodes, N. Gligorov, & A. Schwab (Eds.), *The human microbiome: Ethical, legal and social concerns* (pp. 107–127). Oxford: Oxford University Press.

Greene, J. D., Sommerville, B. R., Nystrom, L. E., Darley, J. M., & Cohen, J. D. (2001). An fMRI investigation of emotional engagement in moral judgment. *Science, 293*, 2105–2108.

Grisso, T., & Appelbaum, P. S. (1995). The MacArthur treatment competency study III: Abilities of patients to consent to psychiatric and medical treatments. *Law and Human Behavior, 19*, 149–174.

HIPAA §164.514(b)(2). (n.d.). *Other requirements relating to uses and disclosures of protected health information*. http://www.gpo.gov/fdsys/pkg/CFR-2002-title45-vol1/pdf/CFR-2002-title45-vol1-sec164-514.pdf. Accessed 3 Oct 2013.

Illes, J., & Racine, E. (2005). Imaging or imagining? A neuroethics challenge informed by genetics. *American Journal of Bioethics, 5*(2), 5–18.

Jackson, F. (1986). What Mary didn't know. *Journal of Philosophy, 83*(5), 291–295.

Kay, K. N., Naselaris, T., Prenger, R. J., & Gallant, J. L. (2008). Identifying natural images from brain activity. *Nature, 452*, 352–355.

Kipnis, K. (2006). A defense of unqualified confidentiality. *The American Journal of Bioethics, 6*(2), 7–18.

Kripke, S. (1980). *Naming and necessity*. Cambridge: Harvard University Press.

Lewis, D. (1990). What experience teaches. In W. G. Lycan (Ed.), *Mind and cognition: A reader*. Oxford: Blackwell.

Logothetis, N. K. (2008). What we can do and what we cannot do with fMRI. *Nature, 453*, 869–878.

Meegan, D. V. (2008). Neuroimaging techniques for memory detection: Scientific, ethical and legal issues. *American Journal of Bioethics, 8*, 9–20.

Nagel, T. (1975). What is it like to be a bat? *The Philosophical Review, 83*, 435–450.

Pardo, M. S., & Patterson, D. (2010). Philosophical foundations of law and neuroscience. *University of Illinois Law Review*, U of Alabama Public Law Research Paper No. 1338763, pp. 1212–1250.

Piccinini, G., & Carver, C. (2011). Integrating psychology and neuroscience: Functional analyses as mechanism sketches. *Synthese, 183*(3), 283–311.

Place, U. T. (1956). Is consciousness a brain process? *British Journal of Psychology, 47*(1), 44–50.

Quine, W. V. O. (1969). Epistemology naturalized. In *Ontological relativity and other essays* (pp. 69–90). New York: Columbia University Press.

Rachels, J. (1975). Why privacy is important. *Philosophy and Public Affairs, 4*(4), 323–333.

Richmond, S. (2012). Brain imaging and the transparency scenario. In S. Richmond, G. Rees, & S. J. L. Edwards (Eds.), *I know what you are thinking: Brain imaging and mental privacy* (pp. 185–203). Oxford: Oxford University Press.

Rorty, R. (1970). Incorrigibility as the mark of the mental. *The Journal of Philosophy, 67*, 399–424.

Roskies, A. L. (2002). Neuroethics for the new millennium. *Neuron, 35*(1), 21–23.

Roskies, A. L. (2008). Neuroimaging and inferential distance. *Neuroethics, 1*, 19–30.

Talbot, M. (2007). Duped: Can brain scans uncover lies? *New Yorker.* http://www.newyorker.com/reporting/2007/07/02/070702fa_fact_talbot. Accessed 6 Apr 2016.

Thompson, J. J. (1975). The right to privacy. *Philosophy and Public Affairs, 4*(4), 295–314.

Tye, M. (1995). A representational theory of pains and their phenomenal character. *Philosophical Perspectives, Vol. 9, AI Connectionism and Philosophical Psychology*, 223–239.

Uttal, W. R. (2001). *The new phrenology: The limits of localizing cognitive processes in the brain.* Cambridge, MA: MIT Press.

Weiller, C., May, A., Sach, M., Buhmann, C., & Rjintjes, M. (2006). Role of functional imaging in neurological disorders. *Journal of Magnetic Resonance Imaging, 23*, 840–850.

Chapter 7
Objectifying Pain

Abstract Pain is characterized as difficult to investigate and to explain using objective scientific means because of its purportedly inherent subjectivity. In this chapter, I distinguish among the various ways in which pain is considered to be a subjective phenomenon, including introspectability, privacy, and incorrigibility. I argue that introspectability and privacy are features that could be shared by states both mental and physical. The kind of subjectivity that is often thought to threaten the scientific study of pain arises only when introspectability and privacy of inner states are coupled with a theory of pain states that posits nonphysical properties to account for the content of pain. Thus, I defend the view that the content of pain can be reduced to what it represents. I also argue that pain is not incorrigible. I argue that the first-person individuation of pain states requires the possession of a rudimentary conceptual framework that includes the concept of pain. This conceptual framework changes over time as an individual is exposed to a variety of different noxious stimuli and acquires a wider vocabulary to express the feeling of pain. Given that the identification of pain requires a concept of pain and that changes in the relevant conceptual framework can alter the feeling of pain, I argue that pain reports are in principle corrigible. Although I acknowledge that there are currently no established criteria to challenge or circumvent the need for a first-person report of pain, I describe some promising new strategies that could lead to the development of such a tool.

7.1 Introduction

On a scale of 1 to 10, with 1 being the mental state easiest to characterize objectively and 10 being the hardest, some philosophers say that pain scores in the double digits. Pain is difficult both to investigate and to explain using objective scientific means because of its purported lack of an appearance and reality distinction. Kripke (1980) captures this feature of pain most aptly when he argues that the appearance property of pain, the way being in pain feels, picks out pain essentially—to feel like you are in pain is to be in pain. Based on this view, the subjective experience of being in pain is what determines the existence of pain for a particular individual. Furthermore, the properties of a pain experience, whether the pain is dull, sharp, searing, and so on, can be characterized only by the person having that pain, i.e., from the first-person perspective. Scientific explanations, however, rely on a third-person perspective, which requires that any properly trained observer could have access to a particular phenomenon, be able to observe it, and assess its properties.

Pain conceived as subjective leads to a further problem, which is that sincere first-person reports of pain are incorrigible. That is, if one regards the subjective experience of pain as the final arbiter of whether a person actually is in pain, then a person in pain cannot be wrong about being in pain. This is unusual, because people are wrong all the time and in a variety of situations when it comes to objective phenomena. Consider perceptual experiences of common objects: A person can be wrong about the properties of objects, such as tables, chairs, apples, and so forth. Even when a distant object appears round to me, it could be square, and my sincere avowal that the object is round is not incompatible with that object's actually being square.

The characterization of pain as subjective is thought of as the commonsense view on pain (Aydede 2005), but it is also adopted by the International Association for the Study of Pain (IASP). The IASP definition of pain states that pain is subjective and that each individual learns the application of the concept of pain based on his or her own individual experiences of pain over a lifetime (IASP 2013). The subjectivity of pain described in this manner is not solely based on the lack of the distinction between appearance and reality; it is also reliant on the known double dissociation between the physical causes of pain—the pain stimulus—and the experience of pain.

A person can feel pain in a limb that does not exist, as in cases of phantom limb pain, or experience a burning pain when stimulated by a dull and cool object (Grahek 2001). Individuals with chronic pain, such as fibromyalgia, experience pain, but no physical cause has yet been discovered to explain it. Moreover, persons with types of brain or nerve damage experience no pain when presented with noxious stimuli, even in situations that result in grave physical damage, such as loss of a finger.[1]

The conclusion that pain is subjective, based on the dissociation between the situations in which a normal individual might feel pain and the experience of pain, requires the assumption that the only candidate for the physical correlate of pain sensation is bodily trauma. However, alternative candidates for the physical correlate of pain experiences are brain states. Imaging studies support this hypothesis because they have been used both to identify neuronal pathways necessary for pain and to correlate individual differences in felt pain with differences in brain activation in particular areas of the brain (Coghill et al. 2003). Moreover, particular kinds of pain, say, pain from different causes, have also been distinguished, using functional Magnetic Resonance Imaging (fMRI) (Baliki et al. 2008). Thus, the dissociation as described by the IASP holds between bodily damage and the experience of pain, but does not hold between instances of brain activity and the experiences of pain. The conclusion, then, that pain must be subjectively defined because of this dissociation is not supported.

---

[1] For some examples, see Grahek (2007 ed.), Chapter 3.

The correlation of brain states with the experience of pain does not, however, resolve the type of subjectivity of pain most emphasized in philosophical approaches to pain. There, the problem is mostly about reducing the qualitative aspects (what it is like to be in a particular state), or qualia, of pain. Some argue that pain can be reduced to what it represents—a disordered state of the body (Tye 1995). Others argue that pain does not have representational content, and is just like an after-image—a purely sensational state with no object (Block 2004).

In Sect. 7.2 of this chapter, I review the scientific theories of pain and discuss the current best scientific models of pain. I focus on the description of the gate control theory of pain, in part to elucidate the now-accepted analysis of pain into its sensory and discriminative, emotional, and cognitive aspects. I also briefly describe some brain imaging studies that localize the distinct elements of the pain system in the brain; such localization is thought to confirm the gate control theory of pain. Although I remain skeptical with regard to the impact of neuroimaging on the philosophical problems associated with the qualitative aspects of pain states, I endorse a representational approach to the content of pain. Such an approach, I argue, is enough to deflate the type of subjectivity of pain described by IASP.

The subjectivity of pain often designates a number of different features, e.g., introspectability, privacy, and incorrigibility. I distinguish among those three and address them separately. In Sect. 7.3, I argue that introspectability and privacy of mental states are obstacles for scientific explanation only if coupled with a view that mental states are nonphysical. A representational view of the content of pain allows for an objective account of pain, even if such states are accessed introspectively and are private inner states. In continuity with my approach in the previous chapters, I rely on the Sellarsian argument that reports of mental states, such as pain, require possession of a conceptual framework that characterizes them and their properties. Using that as a basis, in Sect. 7.4, I argue that first-person avowals are not incorrigible and that there are ways of both influencing and correcting first-person pain reports. How pain is identified and characterized is of consequence both for the scientific study of pain and for how pain is identified and characterized in the medical setting, where treatment of pain is one of the primary obligations of healthcare professionals. Moving away from describing pain as an essentially subjective phenomenon, characterized only through verbal first-person reports, will allow for the utilization of objective means of identifying persons in pain in cases when they are either unable to provide adequate verbal reports or when there is reason to believe their reports are not accurate.

7.2   The Pain System

There are three proposed theories of pain: the specificity theory of pain, the intensity theory of pain, and the gate control theory of pain. Of the three, gate control theory is purported to provide the best account for the different pain phenomena and is currently the dominant scientific account of pain. To highlight the strengths of gate control theory, I will briefly review the two alternative theories.

Specificity theory presumes that there is a dedicated pathway for pain. "The fundamental tenet of the specificity theory is that each modality has a specific receptor and associated sensory fiber... that is sensitive to one specific stimulus" (Dubner et al. 1978).[2] Although Sherrington was the first to describe nociceptors—receptors specifically responsive to noxious stimuli—the existence of those was confirmed later by Burgess and Perl (1967)[3] and Bessou and Perl (1969),[4] who discovered both myelinated and unmyelinated primary afferent fibers responsive only to noxious stimuli. Based on specificity theory, nociceptors relay information received from the periphery of the nervous system to a specific "pain center" in the brain, which was thought to be in the thalamus (Melzack and Casey 1968). A distinction should be made between the terms 'nociceptors' and 'pain receptors': the former is only a claim that there are receptors that respond preferentially to a type of stimulus that might in the right circumstances result in the experience of pain, while the latter assumes the direct relationship between the receptors and the experience of pain. Melzack and Wall accurately characterize this as a psychological assumption (Melzack and Wall 1965, p. 971).

A flaw of specificity theory is the assumption that there is a direct connection between nociceptors and a pain center in the brain. This assumption leaves a number of phenomena unaccounted for, such as pathological pain states that occur in absence of a noxious stimulus, including causalgia, peripheral neuralgia, and phantom limb pain. Causalgia is characterized as a burning pain that may result from a lesion of a peripheral nerve. Peripheral neuralgia is pain felt as the result of peripheral nerve damage. And phantom limb pain is pain experienced as being located in an amputated limb. In each of those cases, the one-to-one relationship between the noxious stimulus and the experience of pain is missing.

There are also pain asymbolia, which occur when the unpleasantness of pain is absent despite the presence of a noxious stimulus. For example, soldiers who, despite being seriously wounded on the battlefield, sometimes do not report being in any pain (Melzack and Wall 1965, p. 150). Patients who have been lobotomized have permanent pain asymbolia. They can appropriately localize the noxious stimuli and rank its intensity, but deny that the stimulus is experienced as unpleasant (Grahek 2007, p. 32). As noxious stimuli and unpleasantness can be dissociated, pleasure and noxious stimuli can sometimes be associated. Such an association was present in the case of Pavlov's dogs, regularly fed immediately after the presentation of a noxious stimulus. Over time, the dogs associated the noxious stimulus with feeding, and during the presentation of the stimulus stopped exhibiting any of the behavior symptomatic of pain (Melzack and Wall 1965, p. 972). Instead, the dogs' reactions to the noxious stimulation were those of joyful anticipation of food. In addition to all these cases of dissociation between pain and noxious stimulation, the specificity approach to pain is challenged by the lack of dedicated pain neurons in

---

[2] Dubner et al. (1978), as cited by Moayedi and Davis (2013).

[3] Burgess and Perl (1967), as cited by Moayedi and Davis (2013).

[4] Bessou and Perl (1969), as cited by Moayedi and Davis (2013).

the central nervous system (CNS), which undermines the notion that there is a dedicated pain center in the brain (Moayedi and Davis 2013, p. 10).

There are two distinct branches of the intensity theory of pain: intensity theory and pattern theory. Intensity theory was championed by Arthur Goldscheider (1984), who claimed that the strength of the stimulus and stimulus summation are the causes of pain rather than the stimulation of dedicated nociceptors.[5] This view was based on studies performed by Bernhard Naunyn in 1859, showing that when non-noxious stimuli were successively applied to the skin, they caused pain in syph-ilitic patients (Moayedi and Davis 2013, p. 8). This was interpreted to indicate that the innocuous stimuli would be summed together until they reached a threshold of strength sufficient to cause pain.

The pattern theory of pain was championed by J. P. Nafe (1929).[6] Based on this view, pain was the result of the activation of a certain spatiotemporal pattern in the brain and not the result of the stimulation of dedicated receptors. "The theory pro-poses that all fiber endings… are alike, so that the pattern for pain is produced by intense stimulation of nonspecific receptors" (Melzack and Wall 1965, p. 973). Sinclair (1955) and Weddell (1955) further showed that the intense stimulation of any fiber would cause pain.[7] Both pattern theory and intensity theory fail on the account that although specificity might not exist at every level of processing for noxious stimuli, there is enough evidence for the specificity of pain receptors and fibers (Melzack and Wall 1965, pp. 971–974).

Many of the pain phenomena not accounted for by either the specificity or the intensity approach to pain can be explained by the gate control theory of pain. The gate control theory of pain was endorsed and described by Melzack and Casey (1968). This theory of pain distinguishes among different elements of pain, which together form the pain system. There is the sensory and discriminative dimension of pain, which is the intensity, location, quality, and duration of pain. The affective and motivational elements of pain include unpleasantness and the flight response to nox-ious stimuli. There are also the cognitive and evaluative aspects of pain, which can mediate the experience of pain based on cultural values, the context in which the stimulus is experienced, and a person's current state of mind. The existing IASP definition of pain adopts this multidimensional description of pain.

The gate control theory of pain requires the postulation of a gate control system that Melzack and Wall (1965) localize in the spinal cord. This gate control system modulates the input transmitted from nociceptors to the transmission cells (T cells) located in the dorsal horn. Afferent nerve fibers or receptor neurons carry informa-tion from the sensory organs to the central nervous system (CNS), and the dorsal horn is where afferent nerves merge into the CNS. The output of the T cells is based on the intensity of the signal from the afferent nerves. The intensity of the output is the ratio of activation between small and large afferent fibers, the latter of which have an inhibitory effect. The large and small fibers are referred to in most of the

---

[5] Goldscheider (1984), as cited in Moayedi and Davis (2013).

[6] Nafe (1929), as cited by Moayedi and Davis (2013).

[7] Sinclair (1955) and Weddell (1955), both as cited by Moayedi and Davis (2013).

literature respectively as C fibers and A-delta fibers (Bishop 1946). The output of the large afferent fibers can also be modulated by the neocortical areas of the brain or by what Melzack and Casey (1968) refer to as the central control system (p. 426).

When the output from the dorsal horn T cells is achieved, it is transmitted towards two distinct brain systems: "a) via neospinothalamic fibers into the ventrobasal and posterolateral thalamus and somatosensory cortex; and b) via medially coursing fibers, that comprise a paramedial ascending system, into the reticular formation and medial intralaminar thalamus and the limbic system" (Melzack and Casey 1968, p. 427). These two distinct, but interacting, brain pathways instantiate the three distinct elements of pain mentioned earlier. The system that projects into the thalamus and the somatosensory cortex underlies the sensory and discriminative aspects of pain, while the activation of the reticular formation and the limbic system contributes to the unpleasantness of pain and motivates the person to perform actions required to avoid noxious stimuli. The third element of pain is instantiated in the neocortical areas of the brain and can mediate the experience of pain.

Melzack and Casey (1968) argue that pain is the result of the interaction of the three dimensions of pain. They argue that no individual brain system should be identified as the pain center in the brain; rather, the entire interacting systems should be taken to be the instantiation of the pain experience. "Pain varies along both sensory-discriminative and motivational-affective dimensions. The magnitude or intensity along these dimensions, moreover, is influenced by cognitive activities, such as evaluation of the seriousness of the injury. If injury or any other noxious input fails to evoke aversive drive, the experience cannot be labeled as pain. Conversely, anxiety and anguish without somatic input is not pain. Pain must be defined in terms of its sensory, motivational, and central control determinants. Pain, we believe, is a function of the interaction of all three determinants, and cannot be ascribed to any one of them" (Melzack and Casey 1968, p. 434).

The complexity of the pain system is interpreted differently by Hardcastle (1999), who argues that the commonsense concept of pain is not representative of the different dimensions of the pain system and therefore should be eliminated and replaced by a number of different concepts that reflect the different elements of the pain system. She also argues that philosophers have erroneously focused on aspects of the pain system and labeled either its affective and motivational facets or sensory and discriminative features as primary and essential for the existence of pain.[8]

Putting these issues aside for now, let us discuss why the gate theory of pain is superior to other theories of pain, based on how it accounts for some of the difficult pain phenomena, including some mentioned in Melzack and Wall (1965). Recall the pain asymbolia exhibited by patients who have been lobotomized (sometimes for the treatment of chronic pain), who report that they still feel pain, but no longer care about it.[9] They are able to report on the sensory-discriminative dimensions of pain,

---

[8] For more, see Hardcastle (1999), Chapter 5, pp. 103–107.

[9] In some cases patients with chronic pain were treated just with a cingulotomy, removal of the cingulate gyrus, which is a brain area that is part of the affective and motivation pathway of the pain system. For some examples, see the Grahek (ed. 2007), Chapter 3.

but no longer feel that the pain is unpleasant. This is because the pathway responsible for the motivational and affective aspects of pain, i.e., unpleasantness, was damaged by the lobotomy. Conversely, damage in the somatosensory cortex has been associated with the loss of the discriminative aspects of the pain experience. Patients with this type of damage report feeling an unpleasant stimulus without being able to locate accurately the source of the feeling or to describe any other aspect of the sensation, for example, whether it is a sharp or a dull pain, or whether it is a burning sensation (Ploner et al. 1999). The different elements of pain described by Melzack and Casey, as well as the description of the different brain pathways, can account for this dissociation in a way that the specificity theory could not because it treated pain as a unitary phenomenon with a single pain center in the brain.

The distinct elements of pain, formulated by Melzack and Casey (1968) have been confirmed using functional magnetic resonance imaging (fMRI). In particular, imaging studies have confirmed that the discriminative and affective aspects of pain are subserved by different brain pathways and that they can dissociate (Rainville 2002). The categories of the affective influence on the experience of pain can be distinguished as pain unpleasantness and suffering, or by what Price calls the "secondary pain effect," which requires the psychological contextualization of pain in terms of its long-term consequences (Price 2000). According to Price, the intensity and unpleasantness of pain are influenced by different psychological factors (Price 2000, p. 1769). But a number of experiments show that it is pain intensity that is the cause of pain unpleasantness. In a study utilizing hypnotic suggestion, modulation of pain unpleasantness affected only that aspect of the felt experience of pain, while hypnotic suggestion targeted to modulated intensity affected both judgments of intensity and of unpleasantness. The secondary effects of pain were shown to be modulated by personality traits, such as neuroticism or extravertism, with neurotics experiencing more suffering as compared with that of extroverts (Harkins et al. 1989). Intensity of pain, however, remained the same across groups, showing that secondary effects of suffering are not based on the sensory and discriminative aspects of pain.

The gate control theory of pain also accounts for inhibitory aspects of the pain system. As described earlier, C fibers in the dorsal column moderate the output of T cells. But the Melzack and Casey model also identifies the neocortical or cognitive inhibitory mechanism. This mechanism can dampen the output from the dorsal horn, preventing the projection of the pain signal into the brain, but it can also inhibit the affective and motivational aspects of the pain system, even after the signal from the T cells has been transmitted through the sensory and affective brain systems.

There is both clinical and physiological evidence that pain can be cognitively moderated as Melzack and Casey described. The physiological evidence shows the effects of anxiety and attention on the felt intensity of pain (Wiech et al. 2008). Hypnosis has been shown to affect the unpleasantness of pain although not the intensity, which is thought to be correlated with the strength of the stimulus (Price 2000). An instance of cognitive mediation of the experience of pain was exhibited by seriously wounded soldiers who denied experiencing any. This was interpreted by Melzack and Wall (1965) to be the result of feeling relieved about having survived. Interestingly, those same soldiers did not lose the ability to feel new, noxious stimuli and would complain about the pain of venipuncture.

Valerie Hardcastle (1999) argues that lack of pain in cases of grave danger can be explained by appealing to the biological role of pain (p. 138–143). The role of pain is to alert the individual of physical injury or the presence of a noxious stimulus in order to activate a flight response. According to Hardcastle, in situations of near death, pain loses its function because there is no escape and the experience of pain is inhibited. She illustrates her claim with a few stories, including one of a man who was almost killed by a lion and yet reported not feeling any pain during the attack. Instead, he described experiencing stupor or dreaminess during which he felt no pain or fear (Hardcastle 1999,p. 139).

Prior learning can modulate the experience of pain. Pavlov's dogs, conditioned to associate noxious stimulation with feeding, did not display any sensation of pain and instead seemed merrily to be anticipating food despite the noxious stimuli that preceded the feeding. Similarly, Hardcastle argues that people involved in self-injurious behavior do not make the usual association between cuts in the skin and the feeling of unpleasantness; instead, they might experience the feeling of relief.

Furthermore, prior painful experience can also affect the experience of pain, illustrating the conclusion by Melzack and Casey that the experience of pain is contextualized within the individual's life-experiences and affects the felt unpleasantness of pain. For example, noxious stimuli perceived to be more life-threatening are ranked as more unpleasant. Interestingly, the perceived threat of the stimulus is based on the individual's assessment of his or her abilities to cope with the threatening stimulus (Wiech et al. 2008).

Hardcastle argues that these inhibitory mechanisms actually comprise a separate system. She proposes a competing theory of pain, which is a modification of the Melzack and Casey (1968) proposal with more emphasis given to the inhibitory aspects of pain. Hardcastle's theory distinguishes between the pain sensory system (PSS) and the pain inhibitory system (PIS) (Hardcastle 1999 130). "PSS and PIS … serve two different goals: the PSS keeps us informed regarding the status of our bodies. It monitors our tissues to maintain their intactness whenever possible. In contrast, the PIS shuts down the PSS when flight or fleeing is imminent, and then enhances the PSS response in moments of calm" (Hardcastle 1999, p. 134). Based on this approach, the PSS is bottom-driven because it is activated by being stimulated noxiously, while the PIS is a top-driven, cognitively controlled system that serves the purpose of modulating the effects of PSS. Hardcastle provides reasons for why she believes that the gate control theory model is in fact two distinct systems, but those reasons are not particularly pertinent here.[10]

---

[10] For more on this, see Hardcastle (1999), Chapter 5.

7.3 The Subjectivity of Pain

Because of the success of the gate control theory model in distinguishing the different aspects of pain, some of the difficult phenomena of pain, including pain asymbolia, can be explained. Nonetheless, the subjectivity of felt experience remains unaccounted for, as it pertains to the affective aspects of pain. Coghill et al. (2003) formulate the subjectivity of pain research in the following way: "an individual's experience of pain, particularly pain of pathological origin, underscores the practical importance of appreciating a first-person experience from a third-person perspective. Undetectable physical differences in injuries or disease processes can result in chronic pain for one individual but only minimal deficit for another. Furthermore, an individual's subjective experience of pain can vary substantially from day-to-day despite being evoked by a temporally invariant stimulus" (Coghill et al. 2003, p. 8538).

The subjectivity of mental states is often used to designate a number of different phenomena that should be distinguished, such as introspectability, privacy, and incorrigibility (Rorty 1970). For now, I will leave incorrigibility aside to address introspectability and privacy, but I will address it in the following section of this chapter.

Mental states are introspectible, which means that they are accessible from the first-person perspective by the person having those states. That they can be accessed in this manner is not proof, however, that they cannot be accessed in any other way or that they are subjective in a way that would lead us to conclude that mental states are nonphysical. One can introspect physical states, including states of the body such as indigestion. Churchland has argued that not just mental states but brain states as well are accessible introspectively (Churchland 1992, 2005). If that is true, then introspectability alone does not lead to subjectivity, because introspectible states can be physical, and those can be studied objectively. Because mental states are accessible from the first-person perspective, they are sometimes said to be private. But if the privacy of mental states designates only the fact that they are inner states, i.e., happening inside the body, that does not make them nonphysical either. Many physical states are private in that way. For example, arrhythmias are inner physical states. Thus, if the subjectivity of pain cited by Coghill et al. (2003) is based on introspectability and privacy alone, it is not an obstacle to scientific explanation, because introspectible and private physical states can be and have been studied scientifically. For example, despite the fact that arrhythmias are introspectible and private, they can be identified and investigated objectively, using a stethoscope or an EKG. Similarly, if mental states are physical states, they could be studied using brain imaging, for example.

Privacy and introspectability can become a problem for scientific explanation when coupled with a view that mental states are not physical, or have nonphysical features, accessible only from the first-person perspective. In what follows, I describe the type of subjectivity that would be an obstacle for scientific explanation.[11]

Consider a person who has a hallucination of a bowl of cherries; this person would if prompted be able to report perceiving a bowl that has a certain color, size, and shape, filled with small, round, and red objects. Given that the person is hallucinating, the properties attributed to the bowl of cherries are true of no physical object currently in front of the observer. In order to explain the experiences of the hallucinating observer, sense-datum theories stipulate a mental object or a sense-datum, of which all perceived properties can be predicated (Aydede 2005, pp. 5–8). Based on the sense-datum view, a person hallucinating a bowl of cherries actually perceives a mental object, not a physical object, with all the properties attributed to it by the perceiver, except that the mental object and its properties are not identical to anything physical in the world. Although the case we described is that of a person who hallucinates a physical object, sense-datum theories argue that mental objects are always present even in cases where perceptual experience has a physical bearer because perceptual experience is always mediated by mental objects that mirror the properties of physical objects. In cases where pain is the result of some obvious damage, such as a stabbed leg, sense-datum theories could argue that the experience of pain is the result of perceiving a mental object that mirrors the physical state of the stabbed leg and has all the properties perceived by the person with the experience of pain. If we turn now to instances where pain is not the result of any obvious physical damage, the sense-datum view applies just as well, because it does not require a physical object to account for the content of the pain experience. The experience of pain is that of the sense-datum of pain.

In some cases, however, pain does not really seem to have an object at all; rather, the sensation of pain is intransitive, with the sufferer attributing properties to the experience itself, not to an object, whether mental or physical. Being in pain is just having the experience of pain. Sense-datum theorists have a good way of handling even such cases by collapsing the act-object distinction and arguing that pain refers both to the object and to the experience (Aydede 2005, p. 8). The object of pain is dependent on the act of experiencing pain. This way of accounting for pain without a physical bearer can certainly account for why a person could experience pain even in cases where no obvious physical tissue damage was present, because the experience of pain would be independent from any objective features of the body. Sense-datum theorists manage to account for pains without physical bearers by making pain a private object accessible for examination and description only by the person experiencing pain. The subjectivity of pain is principled because introspection leads to knowledge of mental, not physical, entities.

To avoid this type of consequence, Pitcher (1970) proposes an alternative view. He argues that the privacy of pain does not have to lead to subjectivity. The content of a state of pain can be reduced to the objective properties represented by that state of pain. The feeling one gets when one is stabbed in the leg is like that of representing

---

[11] For an overview of all views on pain, see Aydede (2005).

the objectively detectable qualities of such physical damage.[12] Pains are like glimpses, according to Pitcher. Although glimpses cannot be distinguished from the act of glimpsing—in order to have a glimpse, there has to be an individual to glimpse it—the privacy of the glimpse is not metaphysical, meaning the facts acquired are not nonphysical. Rather, what is glimpsed are objective states of affairs, such as persons, tables and chairs, cats, and so forth. In the same way as when I have a glimpse of a passing freight train, facts about the content of my glimpse are perfectly objective and they are of the properties of that train. Similarly, my having the experience of pain is required for the pain to exist, but the properties of my experience are objective in the sense that they represent objective states of my body. Thus, that introspection is required for pain to exist is not a problem, because it is used for the representation of objective states of the body. Even though no one but me can feel my pain, if two people were qualitatively sufficiently similar and were exposed to a similar noxious stimulus, their experiences of pain would be similar as well. According to Pitcher, the privacy of pain is not very interesting, because it is not metaphysical subjectivity, which leads to the creation of nonphysical facts.

The representational content of pain is "a token sensory experience which represents that something in the leg is damaged, something moreover that is painful or hurts" (Tye 1995, p. 228). The content of the experience is accounted for by the above representation, even if there is no detectible physical damage to the body. Similarly, when a state has the described representational content, it should be classified as pain even in cases of pain asymbolia where tissue damage does not automatically elicit the feeling of dislike. According to Tye, the categorization of pain and other sensory states happens automatically as the categorization required for vision. A person's visual system can process the direction, location, size, shape, etc., of seen objects without requiring that categorization be the result of conscious thoughts about the visual stimuli. Thus, felt pain can represent the body as being in a certain state without the experiencer's ever having to have conscious thoughts and beliefs about the locus, intensity, sharpness, or any other sensory-discriminative aspect of pain (Tye 1995, pp. 234–237).

That sensory-discriminative aspects of pain are required for an individual to categorize a state as pain is evidenced by a case described by Ploner et al. (1999). The case was of an individual who, due to a localized lesion, lacked any sensory and discriminative aspects of pain. The patient's spontaneous description of the noxious stimuli was as "clearly unpleasant" and located in the area "somewhere between fingertips and shoulder" on his left hand. However, "the fully cooperative and eloquent patient was completely unable to further describe quality, localization and intensity of the perceived stimulus. Suggestions from a given word list containing 'warm,' 'hot,' 'cold,' 'touch,' 'burning,' 'pinprick-like,' 'slight pain,' 'moderate pain,' and 'intense pain' were denied…" (Ploner et al. 1999, p. 213). The patient described in this case had ischemic lesions on the right side of his brain, localized

---

[12] There is some evidence that the commonsense concept of pain is similar to the view described by Pitcher, as people are more likely to think of pain as being located in the affected body part rather than as being a mental particular. For more on this, see Sytsma (2010).

to the cortices SI and SII,[13] and had no other measurable deficits. In addition, his deficits were lateralized to his left side, while his ability to feel noxious stimuli and describe those remained intact on the right side of his body. This case is particularly puzzling because the patient, when stimulated by a noxious stimulus, was not categorizing the sensation as pain, despite being able to identify it as an unpleasant sensation. Unpleasantness is often isolated as the necessary component of pain, but, based on this case, a complete deficit in the ability to identify any sensory and discriminative properties of pain required to form the representational content of pain can result in the inability to experience pain.

Representational accounts of pain, such as Tye's or Pitcher's, ultimately reduce the content of pain to its representational aspects. But as, Block argues, the qualitative aspects of the experience of pain are not representational, i.e., they are not about anything (Block 2004). For example, the experience of a red tomato represents the tomato as red. But the qualitative aspects of the red sensation—what it is like to see red—do not represent anything; they do not attribute any properties to the world. Qualia, according to Block, do not have intentional content; they are not about tomatoes, states of the body, or anything else in the world. If aspects of our bodily sensations do not have intentional content, they cannot be captured by language and are ineffable, according to Block. If they are ineffable, then they are basically not communicable by means of language.[14]

In his attack on functionalism, Block (1980) argued that two individuals who are functional duplicates would be exactly alike except for their qualia. However, he does countenance that the difference in the qualitative aspects of their inner states could be captured by EEG or any other technology that could access brain function as it is happening, such as an fMRI. Aspects of mental states would still remain private because only the person having them would have knowledge of its qualitative features, for example, what it is like to feel a sharp pain. I concur with Tye (2006) and Shoemaker (1981), who contend that for two creatures to be functional duplicates, they must have the same kinds of beliefs about being in pain and the same concept of pain. In addition, our functional doppelgangers will use the same kinds of words to designate pain, including words such as 'qualia' to designate the phenomenal aspects of their experience. Thus, Tye argues, any system that functionally duplicates a person in pain will be in a qualitative state of pain (Tye 2006, p. 159).

But even if one remains unpersuaded by representational accounts of pain, one can still accept my argument against IASP's subjectivist account of pain. The sub-

---

[13] Cortices SI and SII are known to subserve the sensory and discriminative aspects of pain.

[14] Tye distinguishes among representations that have propositional content, those that have sentence-like structures, and those that lack that structure. He argues that the sensory and discriminative aspects of pain represent the body as being in a damaged state in the same way maps represent cities; they are about something, they represent the world or body as having certain properties, but the contents of such representations are not expressible through sentences. Thus, based on Tye's view one could be in an ineffable state that is nonetheless representational. For more see Tye (1995).

jectivity of ineffable qualia is not quite the type of subjectivity expressed in IASP's definition of pain (2013) or cited by Coghill et al. (2003) as an obstacle to scientific inquiry. It is based on the dissociation between the standard circumstances in which pain occurs and the diversity of pain experience reported by individuals in pain. Because this variability in experience of pain has been identified, based on differences in behavior or verbal reports, those differences in felt pain are neither causally inert because they have given rise to beliefs about pain nor ineffable because they resulted in verbal reports.

The subjectivity of pain described by Coghill et al. (2003) is not based on the dissociation between the representational content of pain and the qualitative aspects of pain. It is the more scientifically congenial dissociation between the presence of a particular noxious stimulus and the experience of pain, and the fluctuation of the unpleasantness of pain despite constancy in noxious stimulation. These differences in felt pain, although not correlated with tissue damage or nerve damage, seem to correlate with the activation of the pain system in the brain. For example, differences in sensitivity to pain are not captured by differences in stimuli, but are captured by differences in activation of the relevant brain regions. Coghill et al. (2003), using functional brain imaging, were able to identify interindividual differences in the experience of pain. Their study aimed to investigate objectively the claim that some individuals are more sensitive to pain, experiencing more pain when exposed to a noxious stimulus than other individuals. The subjects in the study were exposed to a stimulus that usually evokes a pain report and were asked to rate the intensity of the stimulus on a scale of 0 to 10. There were individual differences in reported pain intensity, with the most sensitive subjects rating the stimulus as 8.9 out of 10 and the least sensitive subjects rating the stimulus at about 1.05. These differences in pain intensity rankings corresponded with differences in cerebral cortical activation. The most sensitive subjects showed significantly higher frequency of activation in somatosensory cortex, anterior cingulate cortex, and prefrontal cortex, the regions respectively responsible for the sensory-discriminative, affective, and cognitive aspects of pain. There were, however, no significant differences in activation in the thalamus, responsible for the afferent transmission of nociceptive information. The differences in felt pain could not be due to the nociceptive input.

Additional fMRI studies have succeeded in capturing placebo-induced changes in activation in the areas of the brain part of the pain system. Wager et al. (2004) were able to show that the analgesic effect of placebos was correlated with a decreased brain activity in the thalamus, insula, and anterior cingulate cortex. That the placebo altered the experience of pain was evident because the administration of the noxious stimulus was preceded by an anticipation period during which all those areas of the brain had an increase in activity. This study is evidence of the cognitive inhibitory mechanism of pain, where the anticipation of an active analgesic produced an actual relief from pain. But this relief was not merely psychological in the sense that it lacked a physical instantiation; rather, the decreased experience of pain was objectively identified in the brain, using brain imaging.

Derbyshire et al. (2004) conducted an fMRI study comparing the activity in the brain's pain system among individuals experiencing pain in response to noxious stimulation, individuals experiencing pain from hypnotic suggestion, and individuals imagining pain. The brain imaging revealed significant changes during hypnotically induced pain experience within the thalamus and anterior cingulate, insula, prefrontal and parietal cortices. The activated regions corresponded to those usually activated in response to nociceptive stimulation. The intensity ratings differed significantly between the two conditions as did the patterns of activation in the pain system. The analysis of the data also revealed that higher subjective ratings were correlated with higher cerebral activity whether the pain was caused by noxious stimuli or hypnosis.

7.4   The Incorrigibility of Pain

In the previous section, I distinguished between the various ways in which a mental state, including pain, could be subjective. I described introspectability and privacy as two different types of subjectivity. Another way in which the experience of pain could be subjective is by being incorrigible. The incorrigibility of pain reports here does not refer to what the pain is representing. For example, a person having a heart attack might be representing her pain as being in her right arm, when the tissue damage is actually in the heart, thereby misrepresenting the tissue damage as being in the right arm. Incorrigibility stems from the purported lack of an appearance and reality distinction. Rorty defines incorrigibility in the following way: "We have no criteria for setting aside as mistaken first-person contemporaneous reports of thoughts and sensations, whereas we do have criteria for setting aside all reports about everything else" (Rorty 1970, p. 413).

That pain is considered incorrigible is reflected in IASP's insistence that pain be defined as a subjective phenomenon. The IASP definition equalizes what they call psychological pain—pain in the absence of a noxious stimulus—and pain that results from tissue damage. A subjective report of pain is enough for a person to be considered in pain and be treated for it because it is impossible to distinguish between psychological pain and pain caused by noxious stimulation. The IASP definition also contains a reference to how pain enters everyday parlance: "Each individual learns the application of the word through experiences related to injury in early life" (IASP 2013). This part of the definition implies that the ability to report pain is tied to an individual's learning how to use the concept of pain, and given the interpersonal difference in felt pain, each person learns a slightly different concept. It is not entirely clear why the definition limits the learning period of the concept just to childhood, for it is likely that the concept of pain is changed and refined as individuals continue having painful experiences throughout life. In keeping with other chapters of this book, I argue that the concept of pain is required for pain reports, and I use this as a way to argue that pain reports are not incorrigible. I think it is possible to "set aside" a first-person contemporaneous report of pain.

In order to do that, I will draw on Sellars's distinction between *of-ness* of thought and *of-ness* of sensation. For Sellars, Mary's having a sensation of a pink ice cube is a nonconceptual state that Mary is having, while, it seeming to Mary that there is a pink ice cube in the glass is a conceptual state, i.e., a thought that Mary is having. The thought of a pink cube involves concepts like 'pink' or 'cube,' while the sensation of a pink cube does not. Thus, first-person contemporaneous reports of pain are not, based on Sellars's view, sensations but thoughts about pain. To think that reports of pain are reports of sensations is to confuse what is true of the world, or of us (i.e., that we are in a particular state of sensing), and is pre-conceptual with conceptual awareness of things having certain properties, such as being pink or cubed (Sellars 1997 ed. 171).

In order to be able to report having pain, one has to have, at the very least, a rudimentary concept of pain that would enable one to individuate pain as pain and distinguish it from other bodily experiences, such as indigestions, tickles, and twitches. With regard to acquisition of color concepts, Sellars argues: "(T)he process of acquiring the concept of green may—indeed does—involve a long history of acquiring piecemeal habits of response to various objects in various circumstances, there is an important sense in which one has *no* concept pertaining to the observable properties of physical objects in Space and Time unless one has them all…" (Sellars 1977, p. 148). Similarly, the development of the concept of pain requires the learning of the standard circumstances in which pain occurs, exposure to noxious stimuli of different kinds, as well as the acquisition of other concepts required to distinguish among types of pain. In order to be able to distinguish, among sharp, dull, throbbing, burning, and other types of pain, an individual needs to be able to have a thought to the effect that she is in pain and be able to identify a particular kind of pain, i.e., a throbbing pain. Moreover, given that pains are described often metaphorically,[15] descriptions of pain rely on the possession of concepts, such as sharp or searing, which are initially learned in other contexts. Reports of pain likely require a relatively developed conceptual framework. This view of pain is in line with the Melzack and Casey theory. The cognitive mediation of pain experience as described by the gate control theory model gives empirical support for the notion that identification and reports of pain require conceptualizations of sensory states as pains. As was shown in Sect. 7.2, the unpleasantness of pain can vary based on context, mood, and past painful experiences.

Because individuals differ in their life experience, including their exposure to pain, those variations will be mirrored by the conceptual frameworks each of us have. If pain experiences depend on conceptual development, changes in concepts affect the experience of pain. For example, in a clinical setting, a patient will complain to her doctor that she has pain and perhaps report having it in a particular loca-

---

[15] For more on this, see Hardcastle (1999), p. 151.

tion. Since different physical conditions are manifested by different kinds of pain, the doctor will simultaneously attempt to encourage the patient to describe her experience and give her additional vocabulary to render the description more precise. The official diagnostic tool for pain, the McGill Pain Questionnaire, features 20 categories of pain, including descriptions of pain as quivering, lancinating, boring, rasping, as well as descriptions of pain as rhythmic or transient. Although the questionnaire attempts to capture the feeling of pain as experienced by a patient, asking an individual to complete the questionnaire will be likely to develop that individual's concept of pain in a way that enables her to identify a rhythmic, quivering pain—a skill most of us are unlikely to have unless exposed to the questionnaire.

One could imagine a situation in which a person could be persuaded that while she thought she was having a stabbing pain, the pain was actually a drilling pain. For example, a physician attempting to identify the etiology of a particular ache could guide a person to reconceptualize the pain she is having as being of a different kind. The physician could discover that the pain was caused by a pinched nerve and not joint damage, and could point out to the patient that pinched nerves usually result in drilling rather than stabbing pain. If this scenario were real, then it would be the case that the patient who thought she was experiencing one type of pain was mistaken and was in fact experiencing another type of pain. What would likely happen is that the patient would not only be convinced that the pain she was experiencing would be better characterized as drilling rather than stabbing, but she would actually begin to experience the pain as drilling rather than stabbing. The doctor in this case will have corrected the first-person report of a pain state.

Perhaps this argument does not go far enough, because what the doctor corrected was just a mischaracterization of the pain as stabbing. What remains incorrigible is a report that the person is in pain. If pain has no appearance and reality distinction, when it seems to a person that she is in pain, she is in that state. In order, however, for a person to say that it seems to her that she is in pain (the appearance of pain), she would nonetheless have to have a conceptual framework required to individuate pain states. Sellars argues that in order for a person to be able to report that something *seems* to her a certain way requires that she possess a concept of that thing's *being* a certain way. Saying that something seems a certain way is a report on an experience that is, from the first-person perspective, indistinguishable from the experience involved in seeing that something is a certain way for some other consideration, the claim is not being endorsed (Sellars 1997 ed., p. 44). Thus, we have two equal experiences, but when we speak of *seems,* we are withholding endorsement, while when we speak of things being *this* or *that,* we endorse the experience. The concept of something seeming a certain way presupposes the concept of something being a certain way. Moreover, being able to endorse a claim that something is a certain way, presupposes the knowledge of what constitutes standard conditions for detecting such properties, which requires additional concepts. In order for a person to report that it seems to her that she is in pain, she must already possess the concept of pain. Hence, the argument about the potential corrigibility of pain reports would still apply. The appearance of pain depends on the concept of pain and if the concept of pain changes, the appearance of pain would as well. It is not possible for a person to have the experience as of a drilling pain unless she has the requisite concept of drilling pain.

Thus far I focused on the sensory-discriminative aspects of pain. But pain's incorrigibility could stem from its emotional aspects, i.e., pain's unpleasantness. One can be wrong about most of the features of pain, their location, whether they are drilling or stabbing, but not that pain is unpleasant. Moreover, the ability to sense that a state is unpleasant seems not to require concepts. Although it is possible to have an unpleasant sensation without possessing any concepts, it is not possible to have a reportable thought that one is in pain unless one has the requisite conceptual framework. Incorrigibility, however, applies to reports of pain states and requires the conceptual ability to identify oneself as being in such a state. That there is a margin of error for pain reports is evident by the tendency of people to report certain sensations as pain even in cases when they would be best characterized differently. Sensations such as numbness and tingling are sometimes mistaken for pain because they are unpleasant. After being instructed about the differences between numbness, tingling, and pain as well as the situations in which they occur, patients will learn to identify tingling, for example, as distinct from pain.

Although currently there are not accepted third-person criteria that would allow for one to set aside a first-person report of pain, there are reasons to think that such criteria will be developed, some of them perhaps relying on the use of brain imaging. The study by Coghill et al. (2003) presented in the previous section is an example of a way in which such criteria could be established. A certain amount of activity in relevant areas of the brain could be used to determine whether and to what degree a person is in pain. Brain imaging can be used to distinguish between pain and other unpleasant sensations, obviating the need for self-reports. For example, a study by Ploghaus et al. (1999) showed that it is possible to distinguish between anxiety and pain using fMRI alone. The study indicates that regions of the brain active during the period of anticipation of intense pain are different from those active during the experience of noxious stimulus. In another study on pain, Brown et al. (2011) were able to determine whether an individual was experiencing a noxious or an innocuous thermal stimulus, using a support vector machine (SVM) in combination with fMRI. An SVM can be trained on patterns of whole brain activity to distinguish kinds of regional brain activity associated with different types of experiential states (Brown et al. 2011). The whole-brain SVM was used to identify whether an individual was experiencing a painful stimulus. The method proved successful 81 % percent of the time (Brown et al. 2011).

Third-person criteria would be useful in a variety of situations. For example, if patients are not capable of verbal communication, brain imaging could be helpful in determining whether an individual were in pain. The studies cited here, report on individuals who were experiencing pain as a result of noxious stimulation, but this technique is likely to work in nonstandard cases of pain also. For example, Derbyshire et al. (2004) showed that hypnotically induced pain activated similar areas of the brain as for pain in response to a noxious stimulus. Hence, brain imaging could reveal that a person was in pain even if there were no obvious tissue damage. The study by Coghill et al. (2003) indicated that it would be possible to distinguish between the different felt intensities of pain based on brain activity in the pain system.

Although fMRI seem to be the most promising third-person diagnostic tool, other methods have had some success in inferring the presence of pain even in absence of a verbal report. Specifically, Kunz et al. (2007) were able to use facial expressions to determine whether a patient was in pain and even to gauge accurately the intensity of stimulation based on those expressions. In a study about the correlation of self-reports of pain and observed pain behavior, Labus et al. (2003) found a mild correlation between the two, which was most pronounced in acute pain. This method was not as helpful for patients suffering from chronic pain. The participants of the study were individuals able to communicate that they were in pain, but that there is a correlation between self-reports and pain behaviors allows for the possibility for the latter to be used independently.

Brain imaging or any of the other attempts at third-person criteria are not a way of capturing ineffable qualia, but they would be enough to capture a number of features of pain that are most amenable to being expressed in the form of a first-person report. If we take Rorty's formulation of incorrigibility as our guideline, then brain imaging could be developed as a method to set aside some first-person reports of pain. That certain features of pain states can be captured, using brain imaging and other objective methods, does not require the argument that all features of pain states are reducible to brain activity. The subjectivity of pain mentioned by the IASP or by Coghill et al. is not about the subjectivity of ineffable qualia, but about the dissociation between pain and tissue damage. That kind of dissociation, however, is diminished by brain imaging techniques, because they enable us to find a different, more reliable, correlation between felt pain and activation in the pain system. The reliability of this correlation can then lead us to establish criteria that would allow us, in the very least, to corroborate first-person reports of pain. Moreover, it might enable identification of pain states for those patients who are not able to self-report. Even those individuals who are able to communicate their discomfort verbally could benefit from the objective inquiry into the experience of pain because it could lead to a more accurate vocabulary to express pain.

7.5   Conclusion

In this chapter, I have argued against the subjectivity of pain. I distinguish among the ways in which pain states can be said to be subjective. Insofar as a theory of mental states, including pain state, does not posit nonphysical facts or objects, introspectability would not lead to the metaphysical subjectivity of pain states. In line with representational views of pain, I argue that the content of pain states can be reduced to its representational content, which is the representation of the body as being in a disordered state that is unpleasant. Furthermore, I argue that the privacy of pain states is the same as the privacy of all inner states. The dissociation of felt pain from tissue damage leads to the conceptualization of pain as private, but the correlation of pain states with activation in the pain system can be used to diminish the subjectivity of pain. Insofar as there is a correlation between felt pain and activation of brain regions known to instantiate the pain system as described by Melzack and Casey (1968), pain states could at least in principle be identified without relying on first-person reports.

I have also argued against the purported incorrigibility of pain states. I argue that first-person reports of pain states require thoughts to the effect that one is in pain. Thoughts about pain require at least a rudimentary conceptual framework that can allow an individual to identify pain states and be able to report them. This rudimentary framework develops and changes over time as an individual is exposed to a variety of noxious stimuli. In addition, I argue that in order for a person to report that it seems to her that she is in pain—the appearance of pain—she nonetheless has to have a concept of being in pain. I based my argument on Sellars's explication that the concept of being in pain is logical prior to the concept of seeming to be in pain. Finally, I acknowledge that there are currently no accepted third-person criteria to challenge first-person contemporaneous reports of pain. However, I argue that there are reasons to hope that such methods could be developed, using brain imaging. To support my claim, I present several studies in which the development of such criteria has been attempted with some success.

References

Aydede, M. (2005). Introduction. In M. Aydede (Ed.), *Pain: New essays on its nature and the methodology of its study* (pp. 1–58). Cambridge, MA: The MIT Press.

Baliki, M. N., Geha, P. Y., Jabakhanji, R., Harden, N., Schnitzer, T. J., & Apkarian, A. V. (2008). A preliminary fMRI study of analgesic treatment in chronic back pain and knee osteoarthritis. *Molecular Pain, 4*, 47.

Bessou, P., & Perl, E. R. (1969). Response of cutaneous sensory units with unmyelinated fibers to noxious stimuli. *Journal of Neurophysiology, 32*, 1025–1043.

Bishop, G. H. (1946). A-delta fibers are the smallest myelinated fibers, C fibers are the unmyelinated fibers. *Peripheral Nerve Physiological Reviews, 26*, 77.

Block, N. (1980). Are absent qualia impossible? *The Philosophical Review, 89*(2), 257–274.

Block, N. (2004). Mental paint. In M. Hahn & B. Ramberg (Eds.), *Reflections and replies: Essays on the philosophy of Tyler Burge*. Cambridge, MA: MIT Press.

Brown, J. E., Chatterjee, N., Younger, J., & Mackey, S. (2011). Towards a physiology-based measure of pain: Patterns of human brain activity distinguish painful from non-painful thermal simulation. *PloS One, 6*(9), 1–8.

Burgess, P. R., & Perl, E. R. (1967). Myelinated afferent fibers responding specifically to noxious stimulation of the skin. *Journal of Physiology, 190*, 541–562.

Churchland, P. M. (1992). *A neurocomputational perspective*. Cambridge, MA: MIT Press.

Churchland, P. M. (2005). Chimerical colors: Some phenomenological predictions from cognitive neuroscience. *Philosophical Psychology, 18*(5), 527–560.

Coghill, R., McHaffie, J. G., & Yen, Y. F. (2003). Neural correlates of interindividual differences in the subjective experience of pain. *Proceedings of the National Academy of Sciences, 100*(14), 8538–8542.

Derbyshire, S. W., Whalley, M. G., Stegner, A. V., & Oakley, D. A. (2004). Cerebral activation during hypnotically induced and imagined pain. *NeuroImage, 23*, 392–401.

Dubner, R., Sessle, B. J., & Storey, A. T. (1978). *The neural basis of oral and facial function.* New York: Plenum.

Goldscheider, A. (1984). *Ueber den schmerz in physiologischer und klinischer hinsicht: Nach einem vortrage in der Berliner militärärztlichen gesellschaft.* Ann Arbor: University of Michigan Library.

Grahek, N. (2001, 2007 ed). *Feeling pain and being in pain* (Second Edn.). Cambridge, MA: The MIT Press.

Hardcastle, V. (1999). *The myth of pain.* Cambridge, MA: MIT Press.

Harkins, S. W., Price, D. D., & Braith, J. (1989). Effects of extraversion and neuroticism on experimental pain, clinical pain, and illness behavior. *Pain, 36*(2), 209–218.

International Association for the Study of Pain. (2013). Pain Terms. http://www.iasp-pain.org/Content/NavigationMenu/GeneralResourceLinks/PainDefinitions/default.htm. Accessed 12 Sept 2013.

Kripke, S. (1980). *Naming and necessity.* Cambridge, MA: Harvard University Press.

Kunz, M., Scharmann, S., Hemmeter, U., Schepelmann, K., & Lautenbacher, S. (2007). The facial expression of pain in patients with dementia. *Pain, 133*, 221–228.

Labus, J. S., Keefe, F. J., & Jensen, M. P. (2003). Self-reports of pain intensity and direct observations of pain behavior: When are they correlated? *Pain, 102*, 109–124.

McGill Pain Questionnaire. http://www.ama-cmeonline.com/pain_mgmt/pdf/mcgill.pdf. Accessed 12 Sept 2013.

Melzack, R., & Casey, K. L. (1968). Sensory, motivational, and central control determinants of pain: A new conceptual model. In D. R. Kenshalo (Ed.), *The skin sense* (pp. 423–439). Springfield: C.C.Thomas.

Melzack, R., & Wall, P. D. (1965). Pain mechanisms: A new theory. *Science, 150*, 971–979.

Moayedi, M., & Davis, K. D. (2013). Theories of pain: From specificity to gate control. *Journal of Neurophysiology, 109*, 5–12.

Nafe, J. P. (1929). A quantitative theory of feeling. *Journal of General Psychology, 2*, 199–211.

Pitcher, G. (1970). Pain perception. *The Philosophical Review, 79*(3), 368–393.

Ploghaus, A., Tracey, I., Gati, J. S., Clare, S., Menon, R. S., Matthews, P. M., et al. (1999). Dissociating pain from its anticipation in the human brain. *Science, 284*, 1979–1981.

Ploner, M., Freund, H.-J., & Schnitzler, A. (1999). Pain affect without pain sensation in a patient with a postcentral lesion. *Pain, 81*, 211–214.

Price, D. D. (2000). Psychological and neural mechanisms of the affective dimension of pain. *Science, 288*, 1769–1772.

Rainville, P. (2002). Brain mechanisms of pain affect and pain modulation. *Current Opinion in Neurobiology, 12*, 195–204.

Rorty, R. (1970). Incorrigibility as the mark of the mental. *The Journal of Philosophy, 67*(12), 399–424.

Sellars, W. (1977, 1997 ed.). *Empiricism and the philosophy of mind.* Cambridge, MA: Harvard University Press.

Shoemaker, S. (1981). Absent qualia are impossible—A reply to block. *Philosophical Review, 90*, 581–599.

Sinclair, D. C. (1955). Cutaneous sensation and the doctrine of specific energy. *Brain, 78*, 584–614.

Sytsma, J. (2010). Dennett's theory of the folk theory of consciousness. *The Journal of Consciousness Studies, 17*(3–4), 107–130.

Tye, M. (1995). A representational theory of pains and their phenomenal character. *Philosophical Perspectives, Vol. 9, AI Connectionism and Philosophical Psychology*, 223–239.

Tye, M. (2006). Absent qualia and the mind-body problem. *Philosophical Review, 115*(2), 139–168.

Wager, T. D., Rilling, J. K., Smith, E. E., Sokolik, A., Casey, K. L., Davidson, R. J., et al. (2004). Placebo-induced changes in fMRI in the anticipation and experience of pain. *Science, 303,* 1162–1167.

Weddell, G. (1955). Somesthesis and the chemical senses. *Annual Review of Psychology, 6,* 119–136.

Wiech, K., Ploner, M., & Tracey, I. (2008). Neurocognitive aspects of pain perception. *Trends in Cognitive Science, 12*(8), 306–313.

Chapter 8
Identifying Death

Abstract In 1959 two French neurologists, Pierre Mollaret and Maurice Goullon, coined the term *coma dépassé* to designate a state beyond coma. In this state, patients are not only permanently unconscious, but lack brain stem reflexes and the endogenous drive to breathe, indicating that most of their brain has ceased to function. Although legally recognized in many countries as a criterion for death, brain death has not been universally accepted by bioethicists, by the medical community, or by the public. In this paper, I defend brain death as a biological concept. I reassess three assumptions in the brain death literature that have shaped the debate and have stood in the way of an argument for brain death as biological. First, I target the assumption that the biological notion of death has to satisfy a traditional concept of death. I argue instead that the purportedly traditional notion of death is already a scientifically laden concept. Second, I challenge the dualism established in the debate between the body and the brain. Third, I contest the emphasis on consciousness, which prevents the inclusion of psychological phenomena into a biological criterion of death. I propose that the term organism should apply both to the functioning of the body and the brain. I argue that the cessation of the organism as a whole should take into account three elements of integrated function. Those three elements are: (1) the loss of integrated bodily function; (2) the loss of psychophysical integration required for processing of external stimuli and those required for behavior; and, (3) the loss of integrated psychological function, such as memory, learning, attention, and so forth. The loss of those three elements of integrated function is death.

8.1 Introduction

In 1959 two French neurologists, Pierre Mollaret and Maurice Goullon, coined the term *coma dépassé* to designate a state beyond coma. In this state, patients are not only permanently unconscious, but lack brainstem reflexes and the endogenous drive to breathe, indicating that most of their brain has ceased to function. In 1968, an ad hoc committee of the Harvard Medical School, led by Henry Beecher, formulated a brain death criterion of death. Until then, the most widely accepted criterion of death was the complete cessation of circulation of blood and the cessation of vital animal function such as respiration (Beecher 1968). With the advent of medical ventilators designed to move air artificially in and out of lungs, patients whose brain had ceased to function could continue to have a heartbeat for a period of time. Before ventilators, if a whole brain, including cerebrum and brainstem, stopped functioning, the lungs would stop as well because the respiratory cycle depends on a part of the brainstem called medulla oblongata. Lack of oxygen, or hypoxemia, would cause the heart to stop within several minutes. The Harvard ad hoc committee argued that death could be masked by artificial respiration and that an additional criterion of death was required that would utilize diagnostic tests other than those for cardiac death, i.e., circulation and respiration. In the U.S. and worldwide, brain death has since become an established criterion for determining that an individual has died (Wijdicks 2002).

It is noteworthy that the Harvard committee, the 1981 Presidential Commission for the study of ethical problems in medicine, the 2008 President's Council on Bioethics, and other proponents of the whole-brain death criterion have maintained that brain death is not a different kind of death; rather the criterion of brain death refers to the way in which death is determined. Death is defined biologically as the end of the functioning of the organism as a whole, which can be instantiated either as the end of circulation or respiration or, if a person is maintained on a ventilator, as the cessation of whole-brain function. James Bernat, a most persistent defender of the whole-brain death criterion, maintains that death occurs when the brain dies, and that the cessation of circulation and respiration is a good diagnostic test for whole-brain death (Bernat et al. 1981). Formulated in this way, death is, and always has been, the death of the brain.

Despite the insistence that brain death is not an alternative definition of death, or a different type of death, cessation of brain function is still thought of as different from cardiac death. Even some medical professionals still maintain a kind of intuitive distinction between cardiac and brain death, where a brain-dead individual is thought of as not quite dead (Truog 2007). Within the literature on brain death, most commentators, with differing degrees of emphasis, discuss the importance of maintaining a biological definition of death that corresponds to the traditional conception of death. Some of them maintain that brain death could fulfill this traditional notion; others argue that brain death fails to capture it. Whole–brain death is thought to be a technical, science-laden criterion of death, while cardiac death is referred to as the commonsense notion of death—the concept of death a layperson would recognize as an accurate encapsulation of the phenomenon of death. The distinction between the scientific and commonsense notions of death, coupled with the presumption that the commonsense notion ought to be preserved, has led some to argue that brain death in effect needs to fulfill the cardiac criterion of death in order to count as death (Collins 2010; Shewmon 1998). Given that it does not, many have argued that death of the brain is not the death of the organism as a whole (Shewmon 2010).

There are two different types of argument against the criterion of brain death. One argument is that brain death is not death of the organism as a whole because even without a functioning brain, the body maintains its integrity. The brain is not required for the functioning of the organism as a whole. Within this line of argument there have been those who argue that despite the requirement of whole-brain death, clinical tests, which are the only ones necessary for the diagnosis of brain death, cannot show that the entire brain has stopped functioning. Halevy and Brody (1993) argue that the clinical diagnosis of death in fact can miss pockets of functioning areas in the brain. They note that even a proper diagnosis of brain death can miss residual functioning in the pituitary gland, in the cortex, and even some residual functioning of the brainstem. This argument undermines the adequacy of the particular tests performed at the bedside to determine brain death. D. Alan Shewmon, however, presents a more devastating argument against the brain's being the primary organ required for the somatic, or bodily, organization of the human organism. In a series of publications, Shewmon attacks both the argument that cardiac death follows soon after brain death even for individuals with ventilator support and the argument that without the brain the body promptly loses its integrity (Shewmon 1998, 2004). Shewmon argues that, with an intact spinal cord, many of the integrative functions of the body survive the death of the brain (Shewmon 2004).

The second type of argument against whole-brain death as biological is the higher-brain criterion of death. The proponents of this position argue that humans are essentially persons and that the death of the person is the death of the organism as a whole. This view presumes that the concept of a person can be aptly characterized by appeal to certain abilities such as consciousness and other psychological features, most of which are likely to be located in the higher brain, or the cerebrum. Based on the higher-brain death account, the death of the brainstem is not required, because it is irrelevant to the maintenance of the person: patients who are in vegetative states or in a coma are no longer persons and are considered dead. Since the higher-brain death account does not require that the entire brain cease functioning, the lingering pockets of function identified by Halevy and Brody would not be challenging to this criterion. Furthermore, Shewmon's attack on whole-brain death is also not effective against the criterion of higher-brain death because it effectively requires only the end of an individual's psychology, for which bodily integration is not required.

Proponents of the criterion of higher-brain death argue that death is not a biological concept, challenging the notion that death can be defined biologically as the cessation of the functioning of the organism as a whole. Some argue on metaphysical grounds that the death of the person occurs when there is discontinuity of personal identity, which most likely occurs when a human organism loses the ability for consciousness, memory, and other crucial psychological aspects (Green and Wikler 1980; Lizza 2006). Robert Veatch agrees that death cannot be defined biologically; he argues, however, that it is not the end of personhood but the end of embodied consciousness that signals death (Veatch 2005). Furthermore, for Veatch, death is a morally laden concept. The occurrence of death signals the loss of moral standing; those who have died no longer have a claim on our moral regard. Thus, a determination of death can trigger a certain set of death-associated behaviors, such as burial, removal of organs, etc.

In this chapter, I argue that the whole-brain criterion of death instantiates the biological definition of death. The argument is needed because brain death, although legally recognized as one of the two criteria for death, has not been universally accepted as a biological notion on par with cardio-pulmonary death either by bioethicists, by the medical community, or by the public. In the U.S., the less than universal acceptance of brain death as death is evidenced by the legal provisions in some states to allow families who claim moral or religious objections to the brain death criterion to ask for what are called "reasonable accommodations." These accommodations might include a delay in the diagnosis of brain death or a delay of the removal of a ventilator.[1] Furthermore, as has been noted in the literature on brain death, the establishment of brain death as an additional criterion for death has been crucial for cadaveric organ donation, and whether brain death is death in the biological sense can have a significant impact on the permissibility of organ donation after brain death.

To defend whole-brain death, in Sects. 8.2 and 8.3, I describe the traditional notion of death as it appears in the brain-death literature. I challenge the distinction between the traditional or commonsense notion of death as identified with cardiac death, and the technical or scientific notion of death often identified with brain death. I argue that the concept of death, whether commonsense or biological, can change as scientific theories about the biological functioning of the human organism change, and as biomedical advances allow for more ways of extending the life and health of organs critical to life.

In Sect. 8.4, I challenge two assumptions in the brain-death literature that have shaped the debate and have stood in the way of an argument for whole-brain death as biological. The first assumption I challenge is what I characterize as the brain and body dualism prevalent in the debate. The second assumption I contest is the strict association of psychological function with consciousness.

To contest the first assumption, I argue for the expansion of the term 'organism' to include all aspects of brain function. Those who argue against brain death by prioritizing somatic or bodily integration, presume that a biological definition of death (the irreversible cessation of the function of the organism as a whole) may take into account only the integrated functioning of the body either by excluding the brain all together, as Shewmon (1998) does, or including only the functions of the brain that support bodily integration. Based on this view, an individual dies when the body is no longer functioning in an integrated manner, and because the brain is not necessary for somatic integration, brain death would not be death in this sense. Narrowing the term 'organism' in this way, however, to designate only bodily integration, I argue, is not justified and is what prevents the inclusion of all the functions of the brain into the conception of the functioning organism as a whole. I reject the exclusion of brain function from the conception of a functioning organism, and argue that somatic integration is only one aspect of a functioning organism.

To challenge the second assumption, I present evidence for unconscious psychological processes to undermine the strict association of psychological function with consciousness. The second assumption both unduly prioritizes consciousness and

---

[1] In a most recent case in Reno, Nevada, the parents of a young man who had been declared brain dead in St. Mary's Regional Hospital challenged the removal of his ventilator and IV tube. Although a county court ruled that the ventilator and IV tube should be removed, the Nevada Supreme Court in November 2015 overturned that ruling (Klugman 2015).

narrows the scope of psychological function considered relevant for a functioning organism. Moreover, as I argue, for proponents of the higher-brain criteria, the linking of psychological phenomena to consciousness leads to the abandonment of brain death as a biological conception of death.

Finally, I redefine the functioning of the organism as a whole to include three elements of integrated function, without prioritizing any of the elements. The three elements are: the loss of integrated bodily function, i.e., somatic integration; the loss of psychophysical integration required for processing external stimuli and the behavioral outputs of psychological states; and the loss of integrated psychological function, such as memory, learning, attention, etc. The loss of all three elements of integrated function is death of the organism.

8.2   The Different Concepts of Death

Bernat et al. (1981) establish a paradigm for the discussion of death. They maintain that one must settle on the definition of death first. After the definition has been established, one can agree on criteria for death. Currently there are two candidates for such criteria, the cardio-pulmonary criterion that requires the end of circulation and respiration and the whole-brain death criterion that requires the loss of the entire brain. These criteria can be used to specify tests that should be employed at the bedside by physicians to determine whether an individual has died. According to Bernat these levels are discrete and hierarchical; one cannot establish criteria without first settling on a definition, and one cannot settle on right diagnostic tests of death without a proper criterion. Kass (1971) argues for a similar distinction. He argues that settling on the definition of death is a philosophical matter. It is not obvious what Kass means by 'philosophical' in this context, but my assumption is that he is pointing out that the definition of death is underdetermined by the empirical evidence for death. The concept of death is not the same as its physical manifestations that are encapsulated in the criteria for death and determined using tests for death. Because of that, Kass maintains that the discrete aspects of the debate about death require different expertize. Although physicians can diagnose death, they should not be the ones to define it. In this manner, Kass maintains the distinction between the definition of death, and the criteria and tests for death.[2]

In order to settle on a definition of death, many commentators maintain that one must not violate the ordinary meaning of death. Bernat (2002) explicitly states that one should settle on the view of what is commonly meant by death before physicians' measure it (p. 327). Presumably this is a different way of saying that a definition is prior to a criterion of death and the diagnostic tests for death. Moreover, Bernat maintains that death is a nontechnical word and that a definition of death

---

[2] The notion that philosophy is in the business of devising definitions is not contemporary and has been successfully refuted in several different ways in the philosophical literature. Chiong (2005) points this out in relation to the brain death debate (2005, p. 24).

must maintain this aspect of the concept. Other commentators also attempt to capture the ordinary meaning of death. Gert et al. (2006) argue that death is primarily an ordinary term and not one that is medically and legally defined. Therefore, any definition of death, if it is to be accepted, must strive to maintain the ordinary usage of the term 'death.' The 1981 Presidential Commission for the Study of Ethical Problems in Medicine and Biomedical Research similarly emphasizes the importance of not changing the established definition of death. More specifically, the report of the commission states that the cessation of heartbeat and breathing that is recognized as the lay, medical, and legal criterion of death; it is this criterion of death that is meant to exemplify the traditional conception of death.

Based on the literature on brain death, the ordinary concept of death is a concept that persons who are not experts in either law or medicine can hold and use to refer to persons who have died. According to many proponents of the whole-brain death criterion, death ought to be biologically defined as the cessation of the organism as a whole. In response to criticism—from Halevy and Brody (1993), who showed that even with a diagnosis of whole-brain death some patients had intact brain regions, and by Shewmon (1998), who argued that a functioning spinal cord can provide somatic integration—Bernat (2006) amended his definition of death: "Death is the irreversible and permanent loss of the critical functions of the organism as a whole" (Bernat 2006, p. 36). As formulated by Bernat (2006), the critical functions of the organism are (1) consciousness because it is required for the organism to respond to requirements for nutrition and hydration; (2) control of circulation, respiration, and temperature control, needed for the maintenance of cellular metabolism; and (3) integrating and control systems involving chemoreceptors, baroreceptors, and neuroendocrine feedback loops to maintain homeostasis (Bernat 2006, p. 38). Given that the brain is required for all of those critical functions, the death of the entire brain, including the cerebral hemispheres, diencephalon (including thalamus and hypothalamus), and brainstem would be the end of those critical function and thus the individual is dead.

Gert, Culver, and Clouser also characterize death as the permanent end of all observable natural functioning of the organism as a whole, but they add the requirement of the permanent absence of consciousness in the organism as a whole and in any part of the organism (2006, p. 290). They argue that this definition is biological, but that it does capture the traditional notion of death that applies similarly to human beings as to dogs, cats, and other animals. Such a definition is part of U.S. law and is consonant with any number of religious practices. The authors acknowledge that some religions maintain the tenet that there is life after death. They assert that a belief in the afterlife is compatible with the biological definition of death because even the proponents of life after death agree that a living organism has died and that the person persists in a form closely related to but distinct from the original individual. This indicates that even those who believe in the afterlife have adjusted their notions about life and death to accommodate for the biological conception of death when it applies to the human organism.

The requirement that the ordinary meaning of death be preserved entails that the concept of death be both biological and nontechnical. It is not clear how one can achieve either one of those goals. Applying the view I presented in Chap. 2, in order to have a biological concept of death, one must endorse, either explicitly or tacitly, an empirically testable theory that features death among its posited entities.[3] One is able to individuate certain phenomena only insofar as there is a theory that supports that individuation through the attribution of certain kinds of properties. The properties of death are a result of the role of death in a particular theory. Death does not have any properties that are true of it prior to the establishment of the biological theory that features death as one of its entities. Similarly, commonsense concepts, like scientific concepts, are formed through their role in an empirically testable theory. If one accepts my account that commonsense concepts require the endorsement of an empirically testable theory, then any concept of death would be a technical term of a theory. Thus, there are no nontechnical concepts of death, only concepts supported by different theories about the nature of death.

A way in which a commonsense concept of death could be subsumed by a biological concept of death is if the background biological theory of death could reduce the commonsense theory. Such reduction could be obstructed if the commonsense concept and the biological concept of death are part of two incompatible theories each attributing disparate properties to death. Adopting the view that a biological definition should not require any expertise in biology makes this incompatibility most likely because it presumes that a commonsense theory of death is not in any way based on biology. If we construe the ordinary concept of death and the scientific one as being part of two different theories, and if we further require that a commonsense theory is, in Bernat's sense, nontechnical (meaning it does not require endorsement of a biological theory), then we create a permanent incompatibility between the two frameworks. A scientific or biological definition of death cannot account for a commonsense conception of death because that concept is perforce not scientific or biological. One would then have to opt for only one of the two theories.

A way of avoiding such an incompatibility is to omit the requirement that the biological concept of death account for the commonsense concept of death. In previous chapters, I have argued that commonsense conceptions have been influenced by developments in science, and in the next section I will show that this is true of the concept of death. My argument in Chap. 2 is that commonsense conceptions require the endorsement of an empirical theory and that those conceptions change as the theory changes. This is a challenge to the assumption that there is an unchanging traditional view of death that must be accounted for by a biological concept of death. Moreover, if commonsense concepts are construed as outcomes of empirical testable theories, such theories could always turn out to be false and the concepts erroneous. Thus, there is no reason to prioritize a presumed commonsense concept of death.

---

[3] This does not require that any person using the word 'death' have mastery of a biological theory of death or knowledge of the criteria and tests for death, but it does mean that a person using the word death is endorsing a theory about death in the same way as using the phrase "Freudian slip" implicitly commits one to Freud's theory of the unconscious.

8.3 Traditional Death

The report of the ad hoc Harvard committee includes the claim that before the advent of the ventilator, the traditional way of telling that somebody has died is if they stopped breathing and their heart stopped beating. The President's Commission of 1981, however, recounts a time even before the establishment of this cardiopulmonary criterion. During the seventeenth century, doctors were notorious for mistakenly diagnosing death. Coffins were equipped with escape hatches so that those who were erroneously pronounced dead could escape even from the already buried coffin (President's Commission 1981, pp. 12–15). In order to diminish the false positive determinations of death, physicians aimed to perfect their ability to identify the signs of death. Jean-Jacques Winslow argued that the one sure sign of death is putrefaction. In fact, in the nineteenth century, putrefaction was thought of as the only sure way to distinguish apparent death from real death (Carpenter and Gurney 1862). Nowadays, putrefaction is considered part of the process of bodily disintegration that occurs after death (Schmitt et al. 2006). Hence, putrefaction is no longer used to mark the moment of death.

The invention of the stethoscope enabled physicians to determine that a person was dead with more precision because it improved the ability to detect a heartbeat (President's Commission 1981, pp. 12–15). Thus, the moment of death could be associated more precisely to the end of cardiac function since the absence of a heartbeat could be determined accurately. Waiting for putrefaction then became obsolete. Given that before the invention of the stethoscope, the identification of heartbeat in that manner could not have been part of the concept of death, what is now touted as the traditional concept of death is an outcome of scientific discovery. Not only does the concept of death rely on the endorsement of a biological theory of death, it also requires the tacit endorsement of the auxiliary scientific theories that supported the development of the stethoscope and its use to determine the presence and rate of a heartbeat. Thus, the traditional criterion of death actually requires the endorsement of aspects of biology and aspects of other scientific theories that allowed for the development and application of the stethoscope.

The shifts in criteria for death could be interpreted as a refutation of the claim that death is an event. For Bernat and others, including Shewmon, death is defined as a moment rather than a process. Chiong (2005) and Khushf (2010) disagree and argue that death is a process. Those who favor the idea that death is a process claim that the moment of death is just a socially determined point along a continuum there to serve a certain social role in the same manner as turning 18 is determined to be the point at which an individual becomes an adult. Based on the process view, turning 18 does not correspond to any particular biological or physical state whose identification would make a person an adult; rather, it is a relatively arbitrary point in the process of growth and development from adolescence to adulthood, selected to confer upon an individual the rights of an adult.

Those who advocate that death is a moment often favor something akin to a biological definition of death, although they might disagree on what the biological correlates of death are that indicate the moment of death. If death is construed as a process, then this entails that there is not a particular biological event that marks death and therefore it is not possible to argue that the definition of death is biological. Hence, those who argue that death is a process also think that it is our social or moral norms that mark the moment of death. Interestingly, the traditional concept of death is more often characterized as requiring the identification of a particular moment as death. It is thought a challenge to tradition and commonsense to claim that death is not marked by a physical moment, but is constructed based on social norms. Thus, those who are attempting to identify a biological moment of death are the traditionalists in this debate.

There is a way, however, of reconciling the dueling positions, which is to argue both that there is a moment of death and that death is in a certain sense constructed. Based on the view that I endorse, death is posited for the explanation of certain observable phenomena, the properties of death are determined based on the role of death in a background theory, and the moment of death cannot be determined in absence of that theory. It is the endorsement of a particular theoretical view that enables us to identify a particular physical state of the body as death. So it is not that there is a physical state, in absence of a biological theory that makes a difference between life and death: It is the development of a particular biological theory that designates a certain physical state as being the point at which a living organism dies. In this way, one can maintain that death is a physical state of the body and countenance that there are several potential candidates for the moment of death. But it is the endorsement of a particular theory that designates a physical state as death and the moment of death as the occurrence of that state. My view should not be construed as endorsing the position that the moment of death is a social construct with no physical basis. Death is identified by settling on the most explanatorily adequate biological theory about the functioning of the human organism and it is the adequacy of the theory that guarantees the moment of death is identified correctly and the relevant physical state is real.

Developments in biological theory can change the identification of the moment of death. In this section, I have listed two such previous changes. As noted, the discovery of the stethoscope led to replacing putrefaction as the marker of the moment of death. After that discovery, the moment of death was designated as the moment when the heart stopped beating. And with the invention of the ventilator, brain death was identified as an additional physical state that can mark the moment of death. Further biomedical advances could change the moment of death in the future.

Let us evaluate yet another feature attributed to the commonsense conception of death—irreversibility. The 1981 Presidential Commission argues that there are three crucial organ systems of the body—the heart, the lungs, and the brain. The cessation of any of those two can lead quickly to death. Of course many other organs, such as the liver or the kidneys are important for the maintenance of life. Before there were medical ways of either replacing or substituting the function of those organs, the end of their functioning would lead to death. But a nonfunctioning kidney or liver does not mark the death of the human organism because the human can continue functioning for a considerable amount of time after the failure of those organs. Nonetheless, the end of those organs would have in the past been the reliable precursor of death.

The end of the life-preserving triumvirate of lungs, heart, and brain, however, if left unattended, will quickly lead to death. The advent of the respirator now can supplant the functioning of the lungs. Thus, the loss of the lungs does not always lead to the irreversible end of the functioning organism as a whole. The issue of irreversibility has been discussed by David Cole (1992), who argues that there are different ways of thinking of irreversibility. When thinking about irreversible processes as they pertain to the cessation of the functioning of the organism as whole, we can either require that irreversibility cover any conceivable possibility or that it cover only currently available methods that could reverse the demise of an organ. When thinking about a definition of death as being the irreversible cessation of the functioning organism, the sense of irreversibility should apply only to the currently available technology. We are not obligated, for example, to preserve people cryonically in order for them to be revived by some future technology.

Since lungs can be replaced by ventilators, the end of lung function does not lead to the death of the individual and the irreversible end of the functioning of the organism as a whole. The heart also could in some cases be supported either by transplants, pacemakers, or artificial pumps. Thus, the degradation in function of the heart can in some instances be prevented or supplanted and does not have to lead to the death of the organism. The complete loss of neurologic function, however, is currently irreversible. Thus, if brain death is death, then the end of the functioning brain is the end of life. Bernat's (1998) argument with regard to irreversibility goes further, and he claims that the functions of the brain could never be supplanted by artificial processes. It is, of course, true that within the context of current scientific knowledge and technology, it is not possible to replace the functions of the brain. But Bernat's argument is that in principle the brain of an individual could not be artificially supported or replaced. I disagree with this latter aspect of his argument, because it is conceivable, although not yet possible, that an individual's brain could be replaced in a number of different ways while preserving the psychological aspects of that person. If there were a way of replicating the functioning brain of a particular individual, either artificially or through the regeneration of neurons, and, if the memories, personality traits as realized by the brain can be preserved, an individual could survive the death of the brain.

In any case, if the concept of death is tied to irreversibility, it will trail our scientific understanding of irreversible loss of function and it will depend on the ability to replicate the function of the critical organs of the organism. If it becomes possible either to regenerate or to replace basically any organ, irreversibility of organ function will no longer be part of the concept of death, because death would not depend on the irreversible cessation of the functioning of any organ or organ cluster. Therefore, insofar as irreversibility is part of the concept of death, this aspect of the concept is based on a scientific understanding of both the importance of certain organs for the function of the body and our current, scientifically supported, ability to supplant or support the function of those organs.

The reason to emphasize the fact that the traditional concept of death is not commonsense and is influenced by a biological theory is that commentators attacking the concept of whole-brain death do so based on the fact that whole-brain death does not fulfill the criteria of cardiac death. As I have argued, the moment of death is designated by a biological theory, and changes in theory can change the moment of death and as that moment changes, the sequelae of that event should not be expected to be the same. This is important to keep in mind when evaluating Shewmon's attacks on brain death.

Shewmon (1998) describes a number of cases of what he calls chronic brain death. Shewmon documents what he argues is the survival of brain dead patients ranging from 1 week through 7 months. The longest duration a patient was artificially maintained after being diagnosed with brain death was 14 years. Before Shewmon's review of chronic brain death, brain-dead patients were thought to go spontaneously into cardiac arrest, or asystoly, after a very brief period of time. The significant length of time between brain death and spontaneous cardiac arrest, according to Shewmon challenges the notion of whole-brain death. Shewmon points out that in many cases the body of the individual was maintained with little else over and above artificial respiration. This indicates that number of processes crucial for the functioning of the body remained intact despite the demise of the brain. Brain-dead children maintained on respirators continued growing. And there was one reported case of a brain-dead woman who gestated a fetus and gave birth.

Shewmon's argument that chronic brain death is evidence that brain-dead individuals are not dead is not convincing. The definition of death was said to be the cessation of function of the organism as a whole, which can be diagnosed either through brain death or through cardiac arrest. Shewmon's argument seems to rest on the assumption that in order for brain death to be death it should be followed quickly by cardiac death. But the reason for that requirement is not justified. The whole-brain criterion of death is established precisely for cases where the criterion of cardiac death cannot be used to distinguish between live and dead individuals. Thus, it seems peculiar to argue that because cardiac arrest did not proceed quickly from whole–brain death, then brain death is not death. This is especially true if one maintains like Bernat that only brain death is death and loss of cardiopulmonary function should be primarily regarded as a means to diagnose brain death.

Shewmon also seems to presume that the maintenance of somatic function, the function of the body without the brain, should be convincing evidence that death has not occurred. This seems somewhat like arguing that cessation of cardiac function is death only if putrefaction occurs within a certain amount of time. Although one follows usually quickly after the other, we would not say that the cardiac criterion needs to fulfill the criterion of putrefaction in order for cessation of cardiopulmonary function to count as death. Imagine, for example, we freeze a dead body and prevent the process of disintegration; we would not want to argue that the person was not dead because she did not disintegrate. Similarly, it is not clear why the time between brain death and cardiac arrest is relevant to our judgments about the validity of the brain death criterion.

   The fact that aspects of somatic function can be maintained does not mean that the person is alive unless one takes those aspects of bodily functioning to be evidence of life despite the brain death criterion. That a brain-dead body can gestate a fetus, grow, defecate, and retain anti-entropy is not evidence that the person is alive unless one already dismisses the whole-brain criterion of death. Given that the heart and the brain serve two different functions of the body, the cessation of either will affect the functioning of the body in different ways. If we accept that in order for the body to die either the heart has to stop beating or the brain has to die, the cessation of either will count as death. There is no requirement, however, that death in each instance has exactly the same consequences. Consider the example of blindness, where the person could go blind in several different ways. One could in some way lose the functioning of the eye, one could have disruption along the optic nerve, or one could incur brain damage in the visual cortex. In each case the loss of function would result in blindness, but the way in which the person became blind would be different and the way in which blindness would be diagnosed would be distinct as well. Moreover, it would be strange to prioritize either losing function of the eye or the visual cortex or the optic nerve exclusively as the only real type of blindness.

   Analogously, it is possible for the organism as a whole to stop functioning in different ways, and the loss of function would be manifested and diagnosed differently. When a person dies from cardiac arrest, most of the other organs will die within minutes. Nonetheless, pockets of function will remain for a short period of time even after cardiac death. For example, the flaccidity of muscles will continue for a time before they start disintegrating and become rigid; loss of temperature of the body will take some time and can be extremely variable depending on the circumstances of death; the onset of postmortem lividity is also variable (Mclay 2013). But such remaining function is not evidence of life. Hence, if lingering function in the body after cardiac arrest is not enough to compel us to say that person is still alive, then similarly such lingering function should not compel us to say that a brain-dead person is not dead. The fact that such function is more extensive and can continue for a longer period is not enough to conclude that the organism as a whole has not stopped functioning.

8.4   Brain Death

Shewmon's argument against brain death targets the conception of the brain as the integrator of bodily function, the demise of which would lead to the disintegration of the body. Given that death is defined as the cessation of function of the organism as a whole, the moment at which the body stops working in an integrated manner is the moment of death. The disintegration of the body after brain death is thought to be manifested in the following manner—hemodynamic deterioration and loss of homeostasis—which then leads to imminent and irreversible cardiac arrest. Shewmon's argument is that somatic disintegration can be accounted for by spinal shock (Shewmon 2004). Spinal shock can lead to the loss of spinal reflexes as well as to some cardiovascular disorders. But if the patient survives long enough, the spinal shock will subside, and in some cases the individual will recover some normal functioning. This in turn could explain how a brain-dead person who is maintained on a ventilator can retain somatic integration.

Shewmon argues further that brain-dead individuals and persons who have had spinal cord injuries have similar somatic pathophysiology. Shewmon distinguishes between three phases: the induction phase of neurologic lesion; the acute phase; and the chronic phase. Some of the similarities include irreversible apnea, quadriplegia, somatic, and autonomic deactivation, hypothermia and poikilothermia. In the chronic phase after spinal shock has subsided, both populations will recover certain bodily functions including some reflexes, hemodynamic stabilization, and gastrointestinal motility, and even sweating.[4] One of the more notable dissimilarities, however, is that brain-dead patients tend to develop diabetes insipidus, which is absent in patients with spinal cord injury. Despite the similarities between those two populations of individuals, Shewmon argues, one would never say that person who has a high spinal cord injury is dead, even though the integrative functions of the body for both types of individuals are similar.

Shewmon thinks that this is an argument against the biological criterion of brain death. Insofar as the brain death definition of death is dependent on the role of the brain as integrator of bodily function, Shewmon's argument is that the spinal cord can maintain some of that integrative role. If brain death advocates respond by arguing that the remaining somatic integration is not enough to argue that the organism as a whole has maintained function, then they would have to accept that quadriplegics are dead because they have somatic integration similar to that of a brain-dead patients.

Both proponents of brain death, including Bernat, and those who argue that death cannot be biologically defined have taken into account Shewmon's criticism of the brain death criterion. Bernat (2006) has amended his view to accommodate Shewmon's attacks by introducing the caveat that death should be defined as the cessation of all critical functions of the brain, and that the tests for brain death should include more than just a clinical diagnosis (Bernat 2006, p. 35).[5] Others have argued instead that death should be identified with the death of the higher brain, thereby avoiding the importance of somatic integration. If a person is dead when the higher brain dies, then one can account for why quadriplegics would not be considered dead. Persons with spinal cord injury have intact higher-brain function. In addition, one can dismiss the problem of chronic brain death, because brain-dead patients can retain somatic integration for a time after brain death is not important since death occurred when the higher brain died. Most proponents, of higher brain criterion of death, however, abandon the goal of maintaining a biological conception of death that depends on the cessation of function of the organism as an integrated whole.

---

[4] For further details about the similarities and differences between these two kinds of injuries, see Shewmon (2004).

[5] The amendment to the original definition of death has been met with some criticism, characterizing the change as ad hoc, see Collins (2013).

There are a number of proponents of the higher brain criterion of death, and they each argue differently for why the end of function of the higher brain should mark the moment of death. As I stated in the introduction, Veatch argues that that death is not defined biologically; rather it is the end of embodied consciousness. Given that consciousness is thought to be localized in the higher brain, then loss of function of the higher brain is death. Jeff McMahan (1995) argues that there are two kinds of death, the death of the human organism and the death of the person. The death of the human organism can be defined biologically, but the death of the person cannot, even if aspects of the brain are required for the survival of the person: "…the continued existence of the mind, and thus of the self, consists in the survival of enough of the cerebral hemispheres to be capable in principle, or in conjunction with relevant support mechanisms, of generating consciousness and mental activity" (McMahan 1995, p. 107). This is because although the brain is not distinct from the mind (although he argues that the mind and brain are not identical), the brain and the body are distinct, because obviously the brain and the body are not identical. Thus, we end up with a view in which there is the death of the person that occurs when the brain stops supporting consciousness and other aspects of the mind, and there is death of the body, which can be identified with the end of the integrative function of the body.

Green and Wikler (1980) argue that death coincides with the death of the person; their reasons for that are not biological or ethical, but ontological. Insofar as personhood entails the continuity of an individual's psychological features, psychological discontinuity signals the demise of that individual. This view requires that human beings be essentially identified with their psychology. And once brain death occurs, an individual is dead because she no longer has any psychological traits. According to proponents of the higher brain death criterion, the brainstem does not support the psychological aspects of the person, then higher brain death is all that is relevant to our judgment that the individual no longer exists.

Lizza (2006) endorses a pluralistic approach to death, which allows for people to choose a criterion of death based on their religious or philosophical convictions. He argues that a person is a primitive concept, which cannot be analyzed or reduced to anything else, and maintains that the relationship between a human organism and a person is that of constitution, but not of identity. Therefore, it is possible for a person to die before the human organism that constituted that person stops functioning. He argues that death cannot be biologically defined and that death occurs when the psychological integrity of a person breaks down. Lizza (2006) maintains that depending on religious or philosophical views, the point of breakdown differs. He argues that our policies should accommodate this plurality of opinion. Lizza (2006) endorses the view that the end of psychology is the end of a person, which is why he supports the higher-brain criterion of death.

8.4.1 Versions of Dualism in the Brain-Death Debate

There are two versions of dualist approaches in the brain-death debate, mind and body dualism and brain and body dualism. My argument in this section is that the two types of dualism are related in this debate. The exclusion of the brain from the conception of the functioning organism is motivated by the view that inclusion of psychological aspects into a conception of a functioning organism requires endorsement of mind and body dualism. But as I will show, mind and body dualism does not entail brain and body dualism and because of that, brain and body dualism is unsupported.

I will first describe mind and body dualism, which might take the form of substance dualism or property dualism.[6] Substance dualism is the view that there are two substances in the world, the mental substance and the physical substance. Property dualism is the view that although there is only one substance there are two types of properties, physical and nonphysical properties. Based on this view, mental states, which are a type of psychological state, have nonphysical properties. Dualism can be contrasted with physicalism, which is the view that there is only one substance and that all properties are physical properties, including psychological states.

In the brain-death debate there are several proponents of variants of mind and body dualism. Lizza (2006) argues that that the relationship between the human organism and the person is that of constitution, but he also argues that scientific laws about human biology cannot account for all the properties of a person. There are subjective and psychological features of the person that cannot be captured by biology. If we assume that all physical features of a person can be accounted for by a relevant scientific branch, then those features that are not captured by a physical explanation are not physical.

McMahan hesitantly favors a type of dualism (1995). He accepts that a functioning brain is necessary for continuity of personal identity, but he does not accept the view that the mind is just the brain: "If each of us is a substance and each is essentially a mind, then minds are substances, at least in whatever sense in which it is true that you and I are substances. So the mind should not be identified simply with its particular contents. Nor can we say simply that the mind is the brain" (McMahan 1995, p. 103).

Shewmon (2010) is a proponent of Aristotelian hylomorphic dualism. "In this view, the soul is not a spiritual thing like a ghost or an angel that inhabits or is somehow extrinsically related to an essentially mechanical body (substance dualism) but is both the immaterial principle of the intellectual and volitional powers, which operate through the brain but are not reducible to brain activity) and the vital principle, or substantial form, of the body, making it to be precisely a living body" (Shewmon 2010, p. 265). Based on Aristotle's view, the soul is the form and the

---

[6] My brief description of mind and body dualism in this section is included to aid the reader's understanding of this view as it appears in the brain-death literature; a review of the vast literature in philosophy of mind on this topic is outside of the scope of this chapter.

body is the matter, and the combination of the two is required for the existence of a person. Shewmon explicitly states that the soul is immaterial despite its integration with the body, which means that Shewmon is not a physicalist (Shewmon 2010, p. 265).

The second kind of dualism prevalent in the brain-death debate is established between the body, without the brain, and the brain as the generator of psychological function, i.e., brain and body dualism. With Shewmon at the helm of the movement, many of the critics of the brain-death criterion prioritize somatic or bodily integration when it comes to biological death. Proponents of a higher-brain death criterion responded by countenancing that biological death applies only to somatic integration and agree that whole-brain death is not death in the biological sense. To argue for higher-brain death they either prioritize psychological aspects of the individual, arguing that a human being is essentially a person, or they argue, as Veatch does (2005), that death cannot be defined biologically but must take into account the moral dimensions of the event.

The upshot is that the biological definition of death is taken to apply only to the body, not taking into account the brain's role in producing psychological function, and the inclusion of psychological aspects of the person is taken to mean that one is abandoning the biological criterion. This is evident in the following attempts to integrate psychological and biological function. Shewmon (2010) proposes hylomorphic dualism as a way to integrate the two aspects of a person, the psychological and the biological. Veatch (2005) notes the cleft between what could be called the bodily criteria of death and the psychological criteria for death, and he attempts to bridge the gap as well. He argues: "for me…an organism cannot exist as an integrated whole if one of its crucial, essential elements is missing. It is integration of body and mind that is critical, not mere integration of various somatic parts" (Veatch 2005, p. 365). Veatch further argues that the human is more than a body (Veatch 2005, p. 365), which is true if from the concept of body one excludes the brain. But because of the inclusion of the mind into the conception of organism, Veatch abandons the biological conception of death and prioritizes a morally laden one.

Although both types of dualism, mind-body and body-brain dualism, are objectionable, the latter is particularly unmotivated. When surmising the elements necessary for the maintenance of the functioning of the organism as a whole, there is no principled reason to exclude the functions of the brain. Brain-body dualism does not coincide with the dualism between the mind and the body and cannot be supported using arguments underlying that distinction. The justification for mind and body dualism is the argument that mental states are nonphysical and cannot be accounted for by scientific theories that account only for physical properties. Mind-body dualism can be used to distinguish between the physical and the nonphysical properties of an individual, but not between the different physical parts of the individual based on their role in realizing human psychology. The 'body' in mind and body dualism encompasses all physical aspects of the individual, including the brain. Thus, being a mind and body dualist does not require being a brain and body dualist.

If we then reject brain and body dualism as ungrounded, we can argue that the term 'organism' should be used to apply both to the body and to the brain. This means that among the functions of the organism we can include psychological function produced by the brain. This expansion of the term 'organism' is not difficult to support because it does not commit us to a particularly stringent kind of physicalism. More precisely, what is required is the acceptance that a functioning brain is necessary for psychological function, but not the argument that psychology is reducible to neuroscience or other more basic forms of scientific explanation. Moreover, accepting this type of physicalism is incompatible only with types of mind and body dualism that maintain that the mental aspects of an individual can be separated from and maintained without any underlying physical processes, something akin to an argument for an immaterial soul. In order to integrate psychological aspects of a person into the definition of an organism, all that is needed is the commitment to the claim that in order for persons to have psychology, they have to have a functioning brain. The integration of psychological function into the functioning of the organism as a whole does not require abandoning a biological conception of death.

8.4.2   A Broader Construal of Psychological States

For many proponents of the higher-brain criterion of death, consciousness is deemed important for the maintenance of a person, and the irreversible loss of consciousness signals death. Veatch (2005) argues that death is the end of embodied consciousness. Lizza (2006) argues for the importance of the subjective experience for the maintenance of personhood. McMahan (1995) also maintains that death of the person coincides with the loss of capacity for consciousness. Hence, individuals who have lost the capacity for consciousness, such as patients in permanent vegetative states, are considered dead. Even authors who are not proponents of the higher-brain criterion of death identify consciousness as the most important among psychological phenomena. For example, Shewmon argues that biological death, or as he phrases it 'passing away,' coincides with the "permanent absence of both consciousness and circulation of oxygenated blood" (Shewmon 2010, p. 278). Even Bernat numbers consciousness among the critical functions of the organism permanently lost with brain death (Bernat 2006, p. 38).

The cited authors do not provide reasons for prioritizing consciousness over other psychological attributes or mental features. A reason to presume, however, that consciousness is considered essential is that many of the authors take the view that mentality in general requires consciousness. Based on this view, an individual must have the capacity for consciousness to have psychological states. It is also likely that the coupling of consciousness with mentality motivates many of the authors in the debate to opt for mind and body dualism, as was discussed in the previous section.

There are, however, both conceptual and empirical reasons to doubt that consciousness is necessary for mentality. There are philosophers who reject the characterization of mental states as necessarily conscious states and establish the possibility of unconscious mental states. For example, higher-order theories of consciousness distinguish between mental states and conscious states.[7] Based on higher-order views, there is a hierarchy of mental states; there are first-order states, second-order states, and sometimes third-order states. All the states are mental states. But in order for one to become conscious of a first-order mental state, one needs to have a second-order state *about* that first-order state. Similarly, one can become conscious of a second-order state only if one has a third-order state *about* it. For example, imagine I am speaking with a friend, and after five minutes of conversation, I notice he has a mustache. Given that I have been looking at his face for the duration of our conversation, it is likely that I had a number of perceptual, first-order, states about his mustache. But only after a number of minutes do I became conscious of those perceptual states and aware that he has a mustache. I did that by having a second-order state about the first-order perceptual states, about the color, size, direction, and other features of my friend's mustache.

There is also empirical evidence against the notion that psychological states have to be conscious. Experiments utilizing the subliminal prime paradigm have helped demonstrate that perception can be unconscious (Dehaene 2014). For example, multisensory information can be unconsciously coupled, as in the McGurk effect, where visual information of an individual moving their lips to say *ga*, coupled with the auditory stimulus of the syllable *ba*, will produce the conscious perception of the syllable *da* (Dehaene 2014, p. 62). Even further, there is evidence that what becomes conscious is often prescreened by unconscious attention. If attention is conceptualized as a sifter that is required to distinguish relevant from irrelevant information when attending to a task, then there is evidence that attention can operate unconsciously (Dehaene 2014, p. 75). Finally, there is evidence that the unconsciously perceived signals can inhibit automatic responses, an ability previously thought to require consciousness. Participants asked to perform a repetitive task, for example, clicking a key whenever a picture appeared on a screen, were able to inhibit that response when a stop signal was presented. Surprisingly, the stop signal had an inhibitory effect even when it was presented subliminally (Dehaene 2014, p. 85).

The evidence presented here can be taken to challenge even Bernat's view that consciousness is one of the critical functions of the organism because it is required to seek nutrition and hydration (Bernat 2006, p. 38). If consciousness is not always needed for perception, attention, and even halting of automatic behavior, then there is no reason to think that consciousness is required in every instance of seeking nutrition and hydration. I take this not to be an argument against the inclusion of consciousness into the functioning of the organism as a whole, but a reason to expand our conception of psychological function required for nutrition and hydration.

In conclusion, given both the philosophical arguments against the strict association of consciousness with mental states and the empirical evidence that psychological processes can operate subliminally, there is no reason to continue privileging consciousness, as is done in the brain-death debate. It is better to think of conscious-

---

[7] For some renditions of this view, see Rosenthal (1986) and Hill (2005).

ness as one among many psychological processes. Hence, when thinking about brain death, we should not think only about the permanent loss of consciousness, but about the permanent loss of all other psychological abilities typical of humans. A broader construal of psychological states, going beyond the prioritization of consciousness, should increase the amount of relevant aspects constitutive of the integrated functioning of the organism, which leads to the conclusion that the complete loss of the brain results in the loss of functions relevant for somatic integration as well as the loss of all psychological function.

8.4.3 Three Elements of Integrated Functioning

Previously, in Sect. 8.4.1, I argued that the human organism should be construed to encompass both the body and the brain. In Sect. 8.4.2, I argued for the expansion of our conception of psychological function to include not just consciousness, but many more aspects of brain function, such that the demise of brain function should be seen as resulting in the loss of psychological states broadly construed. If the term 'organism' applies both to the functioning of the body and of the brain, then the cessation of function of the organism as a whole should take into account three elements of integrated function: (1) somatic or bodily integration; (2) psychophysical integration; and, (3) psychological integration.

My distinguishing among these three elements mirrors parts of Bernat's formulation of the critical functions of the organism. He formulates the critical functions of the organism in the following way: (1) consciousness, because it is required for the organism to respond to requirements for nutrition and hydration; (2) control of circulation, respiration, and temperature control, needed for the maintenance of cellular metabolism; and (3) integrating and control systems involving chemoreceptors, baroreceptors, and neuroendocrine feedback loops to maintain homeostasis (Bernat 2006, p. 38). In my characterization, critical functions 2 and 3 are joined into the element of somatic integration, while critical function 1 is subsumed under the element of psychophysical integration. In addition to consciousness, which Bernat includes only for its role in nutrition and hydration, I take into account a broader array of psychological processes as crucial for a functioning organism. Unlike others in the debate, I do not prioritize either somatic integration or consciousness, but argue that each of the elements contributes similarly to the functioning of the organism as a whole.

Much of the debate on whether brain death satisfies a biological definition of death centers on the role of the brain in somatic integration. Shewmon (2004) maintains the disintegration of the body after brain death is manifested through hemodynamic deterioration and loss of homeostasis, which then leads to imminent and irreversible cardiac arrest. But he argues that somatic disintegration is the result of spinal shock (not brain death) and if the patient survives long enough, it will subside and, for individuals maintained on ventilators, the spinal cord will maintain somatic integration. According to Shewmon, the brain is not required for somatic integration and brain death is not the death of the organism.

I argue, however, that somatic integration is only one of the elements of the functioning of the organism as a whole. Hence, even if one is persuaded by Shewmon's argument that the spinal cord can maintain somatic integration in brain-dead individuals, this would not be enough to argue that such individuals are not dead. As I argued in Sect. 8.4.1, there is no reason to elevate somatic integration over all other functions of the organism. Based on my view, only the first element of integrated bodily function is somatic integration, which includes "control of circulation, respiration, and temperature control, needed for the maintenance of cellular metabolism; and integrating and control systems involving chemoreceptors, baroreceptors, and neuroendocrine feedback loops to maintain homeostasis (Bernat 2006, p. 38)."

The second element is the psychophysical integration required for processing of external stimuli and the behavioral outputs of psychological states, for example, the kinds of psychophysical integration required to avoid danger or to seek nutrition and hydration. In addition to the loss of integrated bodily functioning, brain death will cause loss of sensory and motor integration required for voluntary behavior. This will result in loss of the integrative brain functioning required for vision, hearing, sense of smell and touch, as well as maintenance of balance. The cessation of brain function will result in the loss of speech comprehension and speech production. Brain-dead individuals lack the ability to represent objects in space and lose the ability to localize their body in space, which results in the inability for spatial behavior. As I argued in Sect. 8.4.2, perception, attention and even inhibitory behavior can operate subliminally. Thus in addition to the permanent loss of the ability for consciousness, brain-dead individuals lack many more psychological abilities required for nutrition and hydration. Thus, I propose that Bernat's critical function 1 be expanded to include all the psychological mechanisms required for psychophysical integration, not just consciousness.

The third element is integrated psychological function required for memory, learning, attention, etc.[8] Brain-dead individuals lack the ability for memory, learning, and a variety of other higher functions, including consciousness and attention. When thinking about the loss of integrated biological functioning of the organism, I maintain that all of those functions of the brain required for psychological and psychophysical integration should be included, in addition to the brain's role in somatic integration.

Conceptions of death are used to distinguish between dead individuals and severely disabled individuals. If one adopts the view that somatic integration is all that counts, then brain-dead individuals might be characterized as severely disabled.

---

[8] My argument does not rely on there being elements of psychological integration that can actually exist in the absence of the abilities to perceive external stimuli and to produce behavior. It might be that most of our psychology requires some degree of psychophysical integration. I contend only that one can distinguish three discrete elements of biological function even if no individual can have integrated psychological function without some degree of psychophysical integration.

If one, however, adopts the view that somatic integration is not enough and that what is required is also psychological and psychophysical integration, then brain death is death. Brain-dead individuals lack psychological integration, psychophysical integration, and aspects of somatic integration. My view does not aim to prioritize either bodily or psychological aspects of a person, because I wish to avoid the complexity required in establishing certain properties as essential for the continuation of human organisms, and others for the survival of persons. All three elements of integrated functioning are similarly required for the functioning of the organism as a whole.

Based on my view, individuals with severe brain injury, who are in a coma, vegetative state, or in a minimally conscious state, should be considered alive. All those states are classified as disorders of consciousness. Patients who are in these conditions have diminished, disturbed, or absent awareness (awareness of self and the environment). Patients in vegetative states and minimally conscious states exhibit diminished arousal (wakefulness) while individuals in coma lack arousal. Arousal depends mostly on areas in the brain stem such as the reticular activating system, while awareness is thought to depend on the functioning of the cerebral cortex. Coma is usually a temporary state, characterized by complete lack of arousal and complete unresponsiveness to stimuli. Individuals who recover partially from coma might become vegetative or minimally conscious. Patients in vegetative states (also referred to as unresponsive wakefulness syndrome) have sleep-wake cycles and exhibit evidence of arousal. Those in minimally conscious states have sleep-wake cycles, show evidence of arousal, as well as evidence of awareness of self and the environment.[9] The tools most frequently used to determine whether an individual is in a vegetative state are clinical history and behavioral observations, and those are not enough to determine whether the individual has lost all cortical function or is only seemingly unresponsive to the environment (Owen et al. 2006). The rate of misdiagnosis for vegetative states is up to 43 % (Owen and Coleman 2008).

Functional MRI (fMRI) has been used to demonstrate that individuals previously thought to be vegetative had some remaining cortical function. Owen et al. (2006) studied a 23-year-old patient who was considered to be in a vegetative state after sustaining a traumatic brain injury in a car accident. In order to test for cortical activity, Owen et al. (2006) conducted an fMRI study during which they gave spoken instruction to the vegetative patient to imagine playing tennis or imagine walking through her house. The fMRI of the vegetative patient showed significant activity in the supplementary motor area while imagining playing tennis. When she was asked to imagine walking through her house, the patient showed significant activity in parahippocampal gyrus, the posterior parietal cortex, and the lateral premotor cortex—areas associated with memory, special organization, and motor function, respectively. The fMRI findings in the patient were indistinguishable from that of a normal person involved in the same mental imagery tasks. The study shows that despite the fact that the patient fulfilled the criteria of vegetative state, she retained the ability to follow spoken command.

---

[9] For more on coma, vegetative states, and minimally conscious states, see Laureys et al. (2004).

A number of additional studies confirmed the usefulness of fMRI in detecting cortical function in patients previously diagnosed as vegetative (Monti et al. 2010, 2013; Naci et al. 2014). A study using electroencephalography (EEG) showed that EEG is better than the clinical exam using only behavioral responses in detecting cortical function and ability to follow commands in patients with severe brain injury. EEG is less costly than fMRI, and that it can be used to detect cortical function in behaviorally unresponsive individuals is encouraging because it is more likely that this technique could be established as a standard diagnostic tool (Cruse et al. 2011). In addition to the evidence of cortical brain function in patients previously diagnosed as vegetative, there is also evidence that zolpidem, a medication approved as treatment for insomnia, can also improve brain function and promote arousal in vegetative patients (Du et al. 2014; Machado et al. 2014). Because of the currently high rate of misdiagnosis for vegetative states, as well as the potential for recovery from those states, it is not warranted to endorse the death of the higher brain as a criterion of death.

Based on my view, even individuals who are accurately diagnosed as being vegetative and show no evidence of cortical activity should be considered disabled, but not dead. In addition to the somatic integration that is retained through the functioning of the spinal cord and functioning of the brain stem, vegetative patients display some psychophysical integration as well. Patients in vegetative states retain the ability to breathe on their own, they have sleep-wake cycles, and they can swallow (Laureys et al. 2004, p. 539). Individuals in minimally conscious states display an even greater level of psychophysical integration because they can follow simple commands, they retain some gestural or verbal responses, and they can produce intelligible speech and even exhibit purposeful behavior (Laureys et al. 2004, p. 539). All of this does not add up to the end of functioning of the organism as a whole, and thus patients in vegetative states and minimally conscious states are not dead, but severely disabled individuals.

8.5   Conclusion

In this chapter, I have argued in defense of the whole-brain criterion of death. I contend that death is not a commonsense concept but that all the properties attributed to death stem from its role in a biological theory about the functioning of a human organism. Thus, a biological concept of death should not be designed to capture any purportedly commonsense notions about death. Furthermore, I note that while a biological theory can establish a physical state as the moment of death, the physical moment of death can change as our theories about human biological function change. I have provided examples of such changes: In the nineteenth century, the moment of death was identified with putrefaction, until the invention of the stethoscope helped establish the cardiac criterion of death. The invention of the ventilator introduced the need for a brain-death criterion and marked complete infarction of the brain as the moment of death. I argue that cessation of cardio-pulmonary function and whole-brain death do not have the same sequelae, and that brain death should not be required to fulfill the cardiac criterion of death.

Additionally, I have argued that a biological definition of death should not be interpreted as referring only to somatic integration. Instead, the conception of cessation of the functioning of the organism as a whole should apply to both the body and the brain. I have maintained that psychological functioning, broadly construed, should be considered relevant for the functioning of the organism as a whole. Finally, I proposed that integrated functioning of the organism as a whole should be redefined to include three elements of functioning: integrated psychological functioning, including memory, consciousness, and emotion; integration of psychological and physical processes, such as responding to fear; and somatic integration. I argue that my redefinition of integrated functioning of the organism as a whole to include three elements of functioning can be used to support the view that brain death is death in the biological sense.

References

Beecher, H. (1968). A definition of irreversible coma. Report of the Ad Hoc Committee of the Harvard Medical School to examine the definition of brain death. *Journal of the American Medical Association, 205*, 337–340.

Bernat, J. L. (1998). A defense of the whole-brain concept of death. *Hastings Center Report, 28*(2), 14–23.

Bernat, J. L. (2002). The biophilosophical basis of whole-brain death. *Social Philosophy and Policy, Summer, 19*(2), 324–42.

Bernat, J. L. (2006). The whole-brain concept of death remains optimum public policy. *Journal of Law, Medicine, and Ethics, 34*, 35–43.

Bernat, J. L., Culver, C. M., & Gert, B. (1981). On the definition and criterion of death. *Annals of Internal Medicine, 94*, 389–394.

Carpenter, W. B., & Gurney, F. (1862). *Principles of human physiology, with their chief applications to psychology, pathology, therapeutics, hygiene, and forensic medicine* (edited with additions). Philadelphia: Blanchard and Lea

Chiong, W. (2005). Brain death without definitions. *Hastings Center Report, 35*, 20–30.

Cole, D. (1992). The reversibility of death. *Journal of Medical Ethics, 18*, 26–30.

Collins, M. (2010). Reevaluating the dead donor rule. *Journal of Medicine and Philosophy, 35*(2), 1–26.

Collins, M. (2013). Brain death, paternalism, and the language of "death". *Kennedy Institute of Ethics Journal, 23*, 53–103.

Cruse, D., Chennu, S., Chatelle, C., Bekinschtein, T. A., Fernández-Espejoet, D., Pickard, J. D., et al. (2011). Bedside detection of awareness in the vegetative state: A cohort study. *Lancet, 378*, 2088–2094.

Dehaene, S. (2014). *Consciousness and the brain: Deciphering how the brain codes our thoughts.* New York: Viking.

Du, B., Shan, A., Zhang, Y., Zhong, X., Chen, D., & Cai, K. (2014). Zolpidem arouses patients in vegetative state after brain injury: Quantitative evaluation and indications. *The American Journal of the Medical Sciences, 347*(3), 178–182.

Gert, B., Culver, C. M., & Clouser, K. D. (2006). *Bioethics: A systematic approach* (2nd ed.). New York: Oxford University Press.

Green, M. B., & Wikler, D. (1980). Brain death and personal identity. *Philosophy & Public Affairs, 9*(2), 105–133.

Halevy, A., & Brody, B. (1993). Brain death: Reconciling definitions, criteria, and tests. *Annals of Internal Medicine, 119*, 519–525.

Hill, C. S. (2005). Ow! The paradox of pain. In M. Aydede (Ed.), *Pain: New essays on its nature and the methodology of its study* (pp. 75–98). Cambridge, MA: MIT Press.

Kass, L. R. (1971). Death as an event: A commentary on Robert Morrison. *Science, 173*, 698–702.

Khushf, G. (2010). A matter of respect: A defense of the dead donor rule and of a whole-brain criterion of determination of death. *Journal of Medicine and Philosophy, 35*, 330–364.

Klugman, C. (2015). The bell tolls for death by neurologic criteria: Aden Hailu. *Bioethics.net Blog.* http://www.bioethics.net/2015/12/the-bell-tolls-for-death-by-neurologic-criteria-aden-hailu/. Accessed 18 Dec 2015.

Laureys, S., Owen, A. M., & Schiff, N. D. (2004). Brain function in coma, vegetative states, and related disorders. *The Lancet, Neurology, 3*, 537–546.

Lizza, J. P. (2006). *Persons, humanity, and the definition of death*. Baltimore: The Johns Hopkins University Press.

Machado, C., Estevez, M., Rodriguez, R., Pérez-Nellar, J., Chinchilla, M., DeFina, P., et al. (2014). Zolpidem arousing effect in persistent vegetative state patients: autonomic, EEG and behavioral assessment. *Current Pharmaceutical Design, 20*, 4185–4202.

Mclay, W. D. S. (2013). Investigation of death. In *Clinical forensic medicine* (pp. 205–217). Cambridge: Cambridge University Press.

McMahan, J. (1995). The metaphysics of brain death. *Bioethics, 9*(2), 91–126.

Monti, M. M., Vanhaudenhuse, A., Boly, M., Pickard, J. D., Tshibanda, L., et al. (2010). Willful modulation of brain activity in disorders of consciousness. *The New England Journal of Medicine, 362*(7), 579–589.

Monti, M. M., Pickard, J. D., & Owen, A. M. (2013). Visual cognition in disorders of consciousness: From V1 to top-down attention. *Human Brain Mapping, 34*, 1245–1253.

Naci, L., Cusack, R., Anello, M., & Owen, A. M. (2014). A common neural code for similar conscious experiences in different individuals. *Proceedings of the National Academy of Sciences, 111*(39), 14277–14282.

Owen, A. M., Coleman, M. R., Boly, M., Davis, M. H., Laureys, S., & Pickard, J. D. (2006). Detecting awareness in the vegetative state. *Science, 313*, 1402.

Owen, A. M., & Coleman, M. R. (2008). Functional neuroimaging of the vegetative state. *Nature Reviews Neuroscience, 9*, 235–243.

President's Commission for the Study of Ethical Problems in Medicine and Biomedical and Behavioral Research. (1981). *Defining death: Medical, legal and ethical issues in the determination of death*. Washington, DC: U.S. Government Printing Office.

President's Council on Bioethics. (2008). *Controversies in the determination of death: A white paper by the president's council on bioethics*. http://bioethics.georgetown.edu/pcbe/reports/death/. Accessed 10 Oct 2013.

Rosenthal, D. M. (1986). Two concepts of consciousness. *Philosophical Studies, 94*(3), 329–359.

Schmitt, A., Cunha, E., & Pinheiro, J. (2006). Decay process of a cadaver. In *Forensic anthropology and medicine complementary sciences from recovery to cause of death*. Totowa: Humana Press.

Shewmon, D. A. (1998). Chronic "brain death": Meta-analysis and conceptual consequences. *Neurology, 51*, 1538–1545.

Shewmon, D. A. (2004). The "critical organ" for the organism as a whole: Lessons from the lowly spinal cord. In O. Machado & D. A. Shewmon (Eds.), *Brain death and disorders of consciousness*. New York: Kluwer Academic/Plenum Publishers.

Shewmon, D. A. (2010). Constructing the death elephant: A synthetic paradigm shift for definition, criteria, and tests for death. *Journal of Medicine and Philosophy, 35*, 256–298.

Truog, R. (2007). Brain death: Too flawed to endure, too ingrained to abandon. *Journal of Law, Medicine, and Ethics, 35*, 273–281.

Veatch, R. M. (2005). The death of whole-brain death: The plague of the disaggregators, somaticists, and mentalists. *Journal of Medicine and Philosophy, 30*, 353–378.

Wijdicks, E. F. (2002). Brain death worldwide: Accepted fact but no global consensus in diagnostic criteria. *Neurology, 58*, 20–25.

Conclusion

The conceptual framework of persons is not something that needs to be reconciled with the scientific image, but rather something to be joined to it. To complete the scientific image we need to enrich it not with more ways of saying what is the case, but with the language of community and individual intentions, so that by construing the actions we intend to do and the circumstances in which we intend to do them in scientific terms, we directly relate the world as conceived by scientific theory to our purposes, and make it our world and no longer an alien appendage to the world in which we are living (Sellars 1991 ed.).

In *Philosophy and the Scientific Image of Man*, Wilfrid Sellars distinguishes between what he calls the manifest and the scientific image of "man-in-the-world" (Sellars ed. 1991, pp. 1–40). This distinction between the images is tempered with the caveat that both are idealizations and that the boundaries between the two are not sharp. The manifest image corresponds roughly with what I have referred to throughout the book as common sense while the scientific image designates science. Based on Sellars's view, the scientific image, although the offspring of the manifest image, is primary, and it is not bound by the categories established by common sense. Interestingly, Sellars states that the manifest image is itself a type of scientific image, which can be refined by the scientific image in ways both empirical and categorical (Sellars ed. 1991, p. 7). The scientific image can refine the already existing categories of the manifest image through revision, but it can also challenge the veracity of the categories of common sense. The contrast between the two images is not, according to Sellars, between the scientific and the unscientific; rather, he argues that the scientific image is distinct because it postulates "imperceptible objects to explain correlation among perceptibles" (Sellars ed. 1991, p. 19).

This particular way of conceiving common sense, both as the basis of the scientific image and as itself a type of scientific image, is what I have used as the impetus for reexamining many of the purported conflicts between commonsense and scientific conceptual frameworks, especially as they pertain to psychology and to morality.

In Chap. 2, I establish, using Churchland's argument, that folk psychology (FP) is a scientific theory, in part because it accomplishes exactly what Sellars states is

© Springer Science+Business Media B.V. Dordrecht 2016
N. Gligorov, *Neuroethics and the Scientific Revision of Common Sense*,
Studies in Brain and Mind 11, DOI 10.1007/978-94-024-0965-9

true of the scientific image: it postulates mental states to explain and predict human behavior (Churchland 1992). Using the Myth of Jones, Sellars argues that Jones discovers that postulating inner states such as thoughts helps him explain and predict the behavior of his compatriots (Sellars 1997 ed.). Only after Jones teaches this theory to his compatriots do they become able to utilize the conceptual framework of mental states. I argue further that the same arguments that apply to folk psychology also apply to folk morality and that it too can be characterized as an empirical theory that we learn to utilize in everyday life.

Recasting folk morality as an empirical theory helps me reexamine the purported conflict between commonsense conceptions of particular moral concepts and the neuroscientific conceptions of the relevant phenomena. I do that by arguing that the boundaries of commonsense morality, as in commonsense psychology, can be circumscribed using David Lewis's method of collecting platitudes from ordinary parlance about a particular domain (Lewis 1972). Collecting quotidian platitudes that feature moral concepts such as free will, personal identity, authenticity, etc. in situations where we are trying to explain and predict people's behavior by referring to these concepts will achieve the functional definition of these concepts based on the causal role specified by the platitudes. The appropriate collection of these quotidian statements, I argue, will include claims derived from scientific theories, such as psychology and neuroscience, in ways that affect the causal role specified by the platitudes and in turn change the definition of the concepts we were seeking to define using Lewis's method. In effect, my argument is both that commonsense morality can be construed as a scientific and empirically evaluable theory and that its ontology has already been influenced by scientific branches like neuroscience.

Using this way of conceptualizing commonsense morality, I have focused on examining the ethical implications of neuroscientific discovery, especially in terms of how they are said to affect our moral concepts, including those of mental privacy, free will, authenticity, the subjectivity of pain, and conceptions of death. There are two strands to the argument, one is anti-eliminativist and the other is anti-essentialist.

Although I have accepted the first step of the eliminativist position, which allowed me to argue for the view that commonsense morality is an empirical theory and that such a theory could be revised by neuroscience, I reject the call for the elimination of our commonsense moral categories. I argue that eliminating commonsense categories requires accepting the incompatibility between commonsense morality and neuroscience in a way that would allow eliminativists to argue that the categories or concepts of commonsense morality are radically false, which I do not think is true. As I have stated before, commonsense moral concepts are not radically false, because they are already influenced by scientific discoveries within neuroscience and scientific psychology. Thus a general incompatibility between the two frameworks does not arise. (Nonetheless, local incompatibilities between the two frameworks do arise and result in ongoing revisions of common sense.) In agreement with Stich, I argue that categorical changes are usually not the result of incompatibilities between the frameworks, but the result of social and political factors (Stich 1996). Moreover, I argue that Lewis's method only allows us to garner a sample of contemporary commonsense morality that is not representative of common sense as such in a way that would allow us to make unqualified claims about the nature of commonsense concepts. Finally, I argue that because of the influence of science on commonsense moral concepts, some of those concepts cannot be properly classified as unscientific.

I adopt this anti-eliminativist stance in Chap. 2, and I also adopt it in Chap. 4, where I argue that Benjamin Libet's characterization of the common notion of free will does not accurately capture common sense (Libet 1999). Thus, I challenge the conclusion that the commonsenses notion of free will should be replaced because of the studies that show that volitional action requires consciousness or the studies that tackle the incompatibility between determinism and free will.

In Chaps. 5 and 6, I have also taken this anti-eliminativist stance when I assess the impact of scientific developments in neuroscience, especially through the development of cognitive enhancers or memory modifiers, on our conception of personal identity. I settle on the concept of narrative identity, which I think is more representative of how we use the concept of identity in everyday parlance, and argue that the use of neuroenhancers or memory modifiers will not challenge our notions of self-identity and psychological continuity. I argue these ways of modifying our psychological traits will become but one way of changing ourselves without threatening psychological continuity. In effect, the utilization of these types of medicines is the realization of Sellars's call for joining the scientific image with the "conceptual framework of persons" and a way of directly relating science to our purposes (Sellars ed. 1991, p. 40). This approach is also embedded in the claims I defend in Chap. 3, namely, that our conceptions of bodily privacy will expand to include the privacy of the information about brain function. I maintain that brain imaging in general, and fMRI in particular, will not challenge what has been described as a folk-psychological notion that mental states are inherently private.

The anti-essentialist strand in my argument relies on Lewis's functionalism as presented in Chap. 2. Based on this view, the definition of theoretical terms is achieved through the specification of their functional role in a background theory. The specification of the functional roles is achieved by collecting the relevant platitudes that refer to particular moral concepts. As platitudes utilized in everyday parlance change over time and as scientific claims become quotidian platitudes, the causal role of the theoretical terms changes and so does their definition. Thus, I argue against the claim that there are certain properties that are essential to the definition of pertinent moral concepts. I do this to dispute the view that commonsense concepts must be the basis for theory development.

This approach is utilized in most of my chapters, but it is most evident in Chaps. 4, 7, and 8. In all three of those chapters, I have argued against the tenet that commonsense concepts could be characterized as having static definitions that ascribe essential properties to them. In Chap. 4, when assessing the argument for the incompatibility of free will with advancements in neuroscience, I challenge the notion that free will is necessarily defined as volitional action that is preceded by conscious intentions. I argue that everyday parlance contains attributions of free will in cases where an individual does not consciously will to act a certain way, most notably in cases of automated action and expert performance in sports. Furthermore, I also show that attributions of free will change to incorporate the biological etiology of certain psychiatric diseases.

In Chap. 7, I have shown that pain need not be defined as a subjective phenomenon—a characterization often attributed to common sense and adopted as such by the International Association for the Study of Pain. And I challenge the claim that first-person reports of pain are incorrigible, a tenet of the view that pain is a subjective phenomenon. Instead, I argue for a more objectivist approach that would allow for the use of third-person methods of ascertaining whether an individual is in pain, such as brain imaging.

In Chap. 8, I defend the argument that the biological notion of death does not have to satisfy the traditional notion of death. The traditional notion of death is described as the unscientific and not-theory-laden concept of death endorsed by the lay public. But I have shown, in this chapter that the traditional notion of death thought to be instantiated by cardiac death, cannot be properly characterized as unscientific. In fact, the cardiac criterion was not the first criterion for death, and was established as a legal and accepted criterion only with the discovery of the stethoscope. I further argue that brain death—established similarly with the advent of a medical device—the mechanical ventilator—should be accepted as biological death without the requirement that it fulfill the traditional notion associated with the cardiac criterion of death.

Joining my two strategies, the anti-eliminativism and the anti-essentialism, I arrive at a picture in which commonsense in general and commonsense moral concepts in particular are continuous rather than incompatible with neuroscience. Commonsense moral frameworks and scientific frameworks, with the relevant domain, are frameworks of the same type as they are empirically evaluable theories, which postulate theoretical terms to account for and to predict observables. Although an empirical theory of its own accord, commonsense morality progresses by absorbing the influences of neuroscience and scientific psychology. These adjustments further minimize the actual incompatibility between commonsense and scientific attempts to capture phenomena within the moral domain. The scientific revision of common sense in this way allows us to more accurately describe and predict human behavior, but it also allows us to promote our morality by utilizing a more accurate picture of human nature. The development of scientific approaches that pertain to human moral abilities should not be seen as the demise of morality, but as a more realistic way of describing our intentions and pursuing whatever our purposes.

References

Churchland, P. M. (1992). *A neurocomputational perspective: The nature of mind and the structure of science*. Cambridge, MA: MIT Press.

Lewis, D. (1972). Psychophysical and theoretical identifications. *Australasian Journal of Philosophy, 50*(3), 207–215.

Libet, B. (1999). Do we have free will? *Journal of Consciousness Studies, 6*(8–9), 47–57.

Sellars, W. (1963, 1991 ed). Science, perception and reality. In *Philosophy and the scientific image of man* (pp. 1–40). Atascadero: Ridgeview Publishing.

Sellars, W. (1977, 1997 ed). *Empiricism and the philosophy of mind.* Cambridge, MA: Harvard University Press.

Stich, S. (1996). *Deconstructing the mind* (Chapter 1, Section 11, pp. 63–91). Oxford University Press.